Creativity in Peripheral Places

Creativity is said to be the fuel of the contemporary economy. Dynamic industries such as film, music, television and design have changed the fortunes of entire cities, from Nashville to Los Angeles, Barcelona to Brisbane and beyond. Yet creativity remains mercurial – it is at the heart of industrial innovation and can attract investment, but it is also an intangible, personal quality and experience. What exactly constitutes creativity? Drawing on examples as diverse as postcard design, classical music, landscape art, tattooing, Aboriginal hip-hop, and rock sculpture, this book seeks to explore and redefine creativity as both economic and cultural phenomenon.

Creativity also has a peculiar geography. Beyond Hollywood, creativity is evident in suburban, rural and remote places – a quotidian, vernacular, eclectic enterprise. In seeking to redefine the creative industries, this book brings together geographers, historians, sociologists, cultural studies scholars and media/communications experts to explore creativity in diverse places outside major cities. These are places that are physically and/or metaphorically remote, are small in population terms, or which because of old industrial legacies are assumed by others to be unsophisticated or marginal in an imaginary geography of creativity. This book reveals the richness and depth, the challenges and surprises of being creative beyond city limits.

This book was originally published as a special issue of *Australian Geographer*.

Chris Gibson is Professor in Human Geography, and currently ARC Future Fellow and Deputy Director of the Australian Centre for Cultural Environmental Research at the University of Wollongong, Australia.

Creativity in Peripheral Places
Redefining the Creative Industries

Edited by
Chris Gibson

Routledge
Taylor & Francis Group

LONDON AND NEW YORK

First published 2012
by Routledge
2 Park Square, Milton Park, Abingdon, Oxfordshire OX14 4RN

Simultaneously published in the USA and Canada
by Routledge
711 Third Avenue, New York, NY 10017

First issued in paperback 2014

Routledge is an imprint of the Taylor and Francis Group, an informa business

British Library Cataloguing in Publication Data
A catalogue record for this book is available from the British Library

ISBN 978-0-415-69668-5 (hbk)
ISBN 978-1-138-79828-1 (pbk)

Typeset in Times New Roman
by Taylor & Francis Books

Disclaimer
The publisher would like to make readers aware that the chapters in this book are referred to as articles as they had been in the special issue. The publisher accepts responsibility for any inconsistencies that may have arisen in the course of preparing this volume for print.

Contents

Introduction—Creative Geographies: tales from the 'margins'

CHRIS GIBSON, *University of Wollongong, New South Wales, Australia*

Much has been written about the geography of creative industries such as film, music, design and fashion, especially in the northern hemisphere. Frequently the focus has been on agglomerations or clusters of activity in districts of major Western cities (e.g. Scott 2006; Bathelt & Gräf 2008; Reimer *et al.* 2008; Watson 2008). Creativity is said to be *the* salient feature of contemporary post-industrial capitalism, fuelling innovation and investment and therefore responsible for urban economic fortunes, as well as being a somewhat intangible quality in places ('the buzz' in urban milieus) responsible for generating lifestyle-led in-migration (Florida 2002). Accordingly, municipal and state authorities in Australia and around the world have rushed to develop strategies aimed at branding places as creative, enhancing creative industry growth and generating the right kinds of built environments to attract and support creative entrepreneurs and firms.

Into this heady mix of speculation on the significance of creative industries, place branding and creative industries policy making, geographers have contributed key insights. Early pioneers were economic geographers such as Allen Scott and Susan Christopherson, who identified industries such as dress-making, film and music as exemplary in relation to debates then raging about the emergence of post-Fordism and its subsequent redistribution of global economic activity (Scott 1984a, b; Christopherson & Storper 1986). Scott's keen instinct to look beyond traditional centres of manufacturing production subsequently led him to be among the first academics to take seriously new 'cultural'[1] industries such as music and film, and examine growing sunbelt cities such as Los Angeles. In Scott's words:

> so much of the literature on urban geography generally was posited on the model of the northeastern US metropolis ... [this] prompted us to attempt to formulate an alternative view based on the very different forms of industrialization and urbanization that were becoming increasingly apparent in Southern California and the rest of the sunbelt. I myself had the extraordinary experience ... of being informed—by an anonymous referee of a research proposal on the manufacturing economy of Los Angeles—that there was no economic activity of this type of any interest in Southern California, and that if you wanted to study urban manufacturing you had better focus on cities like Chicago, Detroit or Pittsburgh. (1999, p. 810)

Examining new industries in previously ignored cities required economic geographers to explore how market logics both similar to and different from traditional manufacturing shaped the geographical distribution of economic activities. On the one hand, new industries such as music, film and fashion were vertically distintegrated, and relied on dense inter-firm transactions. The size, structure and interdependent relationships between creative industry firms encouraged spatial agglomeration in particular districts, usually in large cities: 'agglomeration and related increasing returns effects not only enhance system efficiency but also *creativity*, and perhaps nowhere more so than in the case of cultural-products complexes' (Scott 1999, p. 811). On the other hand, creative production was unstable, unpredictable and continually evolving—meaning that 'frequent access to a large variety of relevant skills is paramount' (1999, p. 812), again encouraging spatial agglomeration.

Accordingly, and as over time the popularity of creative industries spiralled in academia and policy making, a special priority was placed on local cultural activity and face-to-face interactions (Watson 2008). Proximity was important because it underpinned the emergence of what Bourdieu (1993) describes as a 'creative field'—a complex of creative production dependent on an 'industrial atmosphere' (in the Marshallian sense) present in specific urban milieus (Scott 2000). As Leadbeater and Oakley (1999, p. 14) argue,

> Cultural industries are people intensive rather than capital intensive ...
> Cultural entrepreneurs within a city or region tend to be densely interconnected. Cultural entrepreneurs, who often work within networks of collaborators within cities, are a good example of the economics of proximity. They thrive on easy access to local, tacit know-how—a style, a look, a sound—which is not accessible globally.

The theme of big city agglomeration was amplified with the rise in popularity of research-cum-consultancy 'experts' such as Richard Florida and Michael Porter, who toured the world speaking to large public forums and advising governments on the best ways to build creative 'clusters' in their cities (see Gibson & Klocker 2004 for a critical review). Repeatedly promoted were particular aesthetic and cultural qualities of cities—walkability, urban cosmopolitanism, civic support for the arts, diversity, vibrant nightlife—as well as economic development and built environment policies designed to harness the spillover benefits of proximity within designated 'creative districts'. Such districts were often placed by municipal governments in dilapidated ex-industrial neighbourhoods of warehouses as part of 'transition fantasies' hoping to solve social and economic problems through creative reinvention and tourism (Lovering 2001; Barnes *et al.* 2006). Creativity, it seemed, could be cultivated.

Creativity beyond the city?

One effect of this mass of academic work and policy making about creativity has been to shape a particular set of assumptions about where creativity is located, where it is likely to emerge, and why (Brennan-Horley & Gibson 2009). Early choices made by geographers such as Allen Scott to examine industries in cities such as Los Angeles (which, as Scott recalls, were ignored at the time) have since become axiomatic, accompanied by other iconic centres such as Paris and London, and a new suite of cities of creative reinvention subject repeatedly to analysis of both boosterish and critical persuasions, including Manchester and Glasgow, UK, and Barcelona, Spain (Brown *et al.* 2000; Macleod 2002; Young *et al.* 2006; Casellas & Pallares-Barbera 2009; Johnson 2009). It would appear from the literature alone that creative industries have a predilection towards sizeable cities— ex-industrial powerhouses or cities of global stature—and towards particular northern hemisphere cities in the industrialised West.

And yet, as an increasing number of scholars are drawn to point out, this picture is at best very partial. As Sorensen (2009) simply argues, much academic work on creativity displays an unacknowledged urban bias. Researchers have looked for creativity in fairly obvious places (big cities, cities making overt attempts to reinvent themselves through culture, creativity and cosmopolitanism); have found it there; and have theorised about cities, creative industries and urban transformations as if their subsequent models or logic were universally relevant everywhere.

In response, other academics (including Allen Scott himself—writing recently about the English Lake District) have sought to explore how cultural and creative industries emerge from small, suburban, rural and remote places and are implicated in a range of social, economic and technical transformations peculiar to those localities (Kneafsey 2001; Gibson 2002; Gibson & Connell 2004; Eversole 2005; Johnson 2006; Gibson & Brennan-Horley 2006; Markusen 2007; McGregor & Gibson 2009; Waitt & Gibson 2009; Scott forthcoming; Bell & Jayne forthcoming). Just as economic geographers writing about music and film in the 1980s looked to new places and theories, contemporary academic work on creativity has broadened and sought to move beyond a now familiar set of cities where shorthand policy ideas (café culture, 'the buzz', small firms co-located in refurbished warehouse 'creative hubs') have become clichéd.

This special issue is dedicated to further exploring the creative industries outside major cities—in places that are physically and/or metaphorically remote, are small in population terms, or which because of socio-economic status or inherited industrial legacies are assumed by others to be unsophisticated or marginal in an imaginary geography of creativity. In Australia, this agenda is particularly relevant, not just because the 'tyranny of distance' has historically been considered a national economic problem (Battersby & Ewing 2005), but because internally large distances, uneven geographical distribution of wealth, and a polarised urban hierarchy of a few large cities (with mostly suburban populations) and a large number of scattered, small towns characterise the Australian situation. Australia's geography demands creative industries research to look beyond inner-city agglomeration and bohemia—lest research merely identifies a handful of key locations in but a couple of State capitals. This special issue responds to this demand to look elsewhere.

Rethinking proximity

The various articles in this special issue seek to redress urban bias, but also bring into question a range of assumptions and critical issues for the wider study of creative industries. The first of these relates to the importance of proximity in urban milieus. If proximity is vital to the creative field, to enabling face-to-face interactions and reducing risk within volatile markets for creative products ('keeping up with trends'), what challenges does remoteness and smallness generate for creative producers—the fact of being a long way from 'happening' places and scenes?

Gibson *et al.* (this issue) explore how in Darwin, challenges of remoteness include immediate issues such as the costs associated with flying in performers and artists for festivals and exhibitions, and more complex issues such as the difficulties of accessing key industry gatekeepers. In Broken Hill, distance from cities and important events was far and away the key locational disadvantage identified in a survey of creative professionals (Andersen, this issue). Remoteness 'means limited types of creative making; wariness of newcomers and new ideas; the loss of young people; limited access to business expertise, production services and training; lack of cultural stimulation; and high transport costs'. Similar problems have faced Wangaratta (Curtis, this issue), Glen Innes (Connell and Rugendyke, this issue) and other regional towns. Loss of young people also resonates in Andrew Warren and Rob Evitt's examination of Indigenous hip-hop musicians from the two disparate locations of the Torres Strait Islands and Nowra, New South Wales. In both cases, despite their different proximities to major cities, young Indigenous performers felt they needed to move to improve career chances. In Perth (Bennett, this issue), classical musicians considering changing orchestras did so in the knowledge that it would mean moving States—and thus involving massive distances. Even in the outer suburbs of a large city like Brisbane (Felton *et al.*, this issue), this sense of distance from key sites of interaction and decision making can be present, encouraging new forms of networking and interconnectivity.

Centripetal forces

A second theme concerns relationships between small and remote places and their larger counterparts, especially those cities where power centralises within the creative industries. In a parallel to the position of Australia in wider global markets (Battersby & Ewing 2005), small and remote places run the risk of perennial peripherality—especially when attitudes to certain places as 'culturally arid' persist (Bennett; Felton *et al.*, this issue). Such places have 'image problems', and lose talented up-and-comers in the creative industries to larger centres—a centripetal trend not only present in creative fields but characteristic of wider economic and demographic flows from country towns, industrial cities, and broadacre farming regions (Gabriel 2002). Bennett (this issue) explores centripetal forces in the context of creative workers migrating from Perth in pursuit of career opportunities elsewhere. Beyond the youth out-migration typical of rural regions and even cities like Perth (see Davies 2008), creative workers move for 'pilgrimage' and career development, often temporarily, and many never return. Bennett traces the experiences of those who have left, documenting the ambivalence of Perth's creative diaspora towards their home town. In the places left behind the result is lamentation for the loss of talent and leadership within local scenes. All this is set against a context in which Perth itself is the centre of intense centripetal forces within Western Australia, dominating that State's creative industries more than any other capital city in Australia. By contrast, in Verdich's analysis of the migration motivations of the so-called 'creative class' to Launceston, Tasmania, creativity played little or no role in attracting people—instead migrations were motivated by the lure of Launceston's 'smallness' (and promises of a quieter life), or the prosaic desire to be nearer family and loved ones (also a high priority for Felton *et al.*'s suburban creative workers, especially as they grew older; cf. Bell & Jayne 2006; Hansen & Niedomysl 2009). In Launceston, creativity only became important once recently arrived migrants realised the town's comparative cultural wealth—then encouraging people to stay.

Negotiating marginality

Leadbeater and Oakley's (1999, p. 14) argument is that with local know-how and skill, places can negotiate their marginal position in relation to global cultural and economic flows, and 'sell into much larger markets but rely upon a distinctive and defensible local base' (see also Power & Jannson 2008). Earlier experiences in the music industry demonstrated how this might be possible: the examples of the 'Dunedin Sound', and Icelandic alternative rock demonstrate how remoteness can be woven into claims about distinctiveness and quirkiness (said to be a product of isolation) as they are marketed globally (McLeay 1994; Connell & Gibson 2003; Stratton 2008). Tourism, of course, also enables connections between small, remote or inaccessible places and global markets (indeed, tourism often *trades* on remoteness, as exoticism). This, too, creates opportunities for creative producers.

In this special issue several examples prominently illustrate this. Gibson *et al.* discuss ways in which remoteness was considered by many to be a delight, bringing solitude and freedom from metropolitan whims and fashions. Accordingly, tropicality and Indigeneity infuse the marketing and critical reception of much creative work from Darwin. Indeed, in the trade for Aboriginal art and music, remoteness is critical for securing the authenticity required to successfully market products (Myers 2002; Gibson & Connell 2004; Christen 2008). Andersen explores how iconic outback landscape, cheap rents, reliable tourism influxes, desires for solitude and the pioneer work undertaken by key individuals laid the foundations for a vernacular visual art industry in Broken Hill, in landscape tradition, to emerge outside the metropolitan-dominated Australian art scene. Ignored by big city critics and dealers, a sense of local distinctiveness, rooted in landscape, sells Broken Hill art.

Perhaps in the tourism industry, above all, remoteness, marginality and difference can be brokered into a base for a distinctive and successful industry. Connell and Rugendyke trace the creation of a Celtic countryside (even a theme park) in the northern New South Wales town of Glen Innes, fixed in place by the construction of the massive Standing Stones. By focusing and embellishing one strand of history (and denying others) tourism has been stimulated in what would otherwise have been a regional centre of no great consequence. Even more quirky examples of creativity, such as the remarkable Utes in the Paddock at Burrawang in central New South Wales (see www.utesinthepaddock.com.au) have drawn tourists to otherwise unpropitious locations.

Somewhat similarly, Curtis explores how the Wangaratta Festival of Jazz has been able to establish itself in an otherwise little-known town as the country's premiere jazz event, by maintaining a commitment to musical experimentation (and ceding creative control over to the musicians themselves). Centripetal forces are inverted at the time of the festival when metropolitan musicians descend on Wangaratta to an event and town that is revered within that creative community. The event has maintained credibility and sustained success (not always an easy combination) by foregrounding musical creativity, as defined by jazz musicians, as its most important priority. Meanwhile, in Mayes' analysis of postcard production in rural Ravensthorpe, Western Australia, distance and smallness meant that national postcard production companies simply ignored this place—creating a vacuum that locals were keen to fill in their own modest, but socially rich way.

In other examples explored in this special issue, the means to negotiate marginality are not nearly so simple. People living and working in remote, rural and small places manage and contest their geographical position or perceived marginality on a daily basis—not always reconciling the pleasures of isolation with the limitations of distance. As Warren and Evitt explore, for Indigenous hip-hop musicians in the Torres Strait new telecommunications and recording technologies have helped counteract difficulties of being remote from key creative centres. Music is recorded using free software, often on communally provided computer equipment, and tracks are uploaded onto MySpace and YouTube for distribution. Tens of thousands of 'hits' online attest to the ability of hip-hop musicians from remote islands to find wider audiences. And yet distance, in combination with cultural norms about what constitutes 'authentic' Indigenous creative expression, continues to circumscribe paid performance opportunities. Indigenous hip-hop musicians rely on gigs at national indigenous festivals and at events celebrating indigenous culture—but struggle to secure slots on 'mainstream' festival line-ups. For one hip-hop group from Nowra on the New South Wales south coast, the only paid gigs were those where raps and rhymes were replaced by loin cloths, didjeridus and 'traditional' Aboriginal dancing—a constraint on creativity framed by dominant (often tourist) understandings of Aboriginality (see also Connell & Gibson 2009). In a sense, then, remoteness can be less a physical geographical and more an imaginary or discursive condition.

What is creativity—and to what purposes is it put?

In the wider literature on creative industries, debate continues on what 'creativity' actually is: whether constant dynamism or vernacular expressivity; pure novelty or 'innovation' within new (predominantly digital) forms of capitalist production; a capacity inherent in everyone (but just waiting to be harnessed through mentoring and inspiration); a convenient term (if indefinable) for a range of employment types in industries in which various forms of cultural expression are commodified for sale, or an 'ideal' quality that policy makers seek to bring out in individuals who are otherwise 'problem' subjects (the unemployed, sick, etc.) (Osborne 2003; Gibson & Klocker 2005; Higgs & Cunningham 2007; Markusen *et al.* 2008; Edensor *et al.* 2009; O'Connor 2009).

The articles collected here in various ways contest dominant ideas of what 'creativity' might be in the creative industries. They discuss what counts as creativity within small, remote and rural places—those places assumed by others to be 'uncreative' because of histories of farming or manufacturing. The stories are often far from straightforward, suggesting the polysemic nature of 'creativity' and the multiple purposes to which creativity is put. In Darwin, conventional creative industries are present, but small, and activities as diverse as whip-making, tattooing and gardening are part of the creative economy. In Wangaratta, genre-specific ideas about experimentation, improvisation, proficiency and spontaneity govern what is deemed 'creative' in jazz, and are pivotal to the credibility of its annual jazz festival (Curtis, this issue). This credibility can only be maintained (and thus the monetary success of the festival assured) when, somewhat paradoxically, the musical creativity present is kept *anti*-commercial.

In Aboriginal hip-hop, proficiency is disavowed in favour of raw re-telling of streetlife stories in rap format—even by those with little or no prior experience or expertise (Warren & Evitt). Expression is creative when considered oppositional, confrontational and political, even when commercial. In Mayes' analysis, the work of producing postcards was indeed creative, but Romantic notions of the individualistic creative self were absent. Postcard producers were self-effacing, and creativity was subservient to the social goals underpinning the activity, 'suggestive of creativity as a means to enhance interaction rather than ("just") interaction as a means to enhance creativity'. Put simply, creativity was a means to an end—that end being the desire to make community, to connect with other people in a remote, lonely place. If in research conducted in large cities creativity has been shown to be contradictory and elusive, then a sense of incongruity is only further intensified beyond the metropolis.

Doing creative industries research differently

Finally, creative industries research very much bears the imprint of the contexts from which it stems (Gibson & Klocker 2004). More than just filling gaps about where creativity might exist beyond big cities, this special issue presents an opportunity to explore how creative industries research can be done differently. Exploring creative industries in rural and remote places, in socio-economically disadvantaged and suburban places, means researchers cannot take context for granted, unlike in cities where urbanity is a given. Warren and Evitt show how, for instance, remote creative industries research necessitates engagements with postcolonial theory: patron discourses fashioned by global metropolitan audiences frame what is possible from Indigenous creative artists. Brennan-Horley, Felton *et al.* and Bennett each in different ways show how creativity remains 'hidden' in small, marginal and remote places that don't have obvious hubs or creative districts, requiring new methods (such as Brennan-Horley's ethnographic mapping approach) and intrepid snowballing (in the cases of Bennett and Felton *et al.*) to identify and then understand the geography of hidden, scattered creativity. As Andersen argues, population churn, itinerancy and reliance on a few key individuals means that creative industry research in small, rural places requires an almost archaeological approach to uncovering 'forgotten' knowledge.

Ultimately, what is the specific significance of creativity to everyday life beyond the big city? Warren and Evitt describe how creativity was a means to self-determination and political expression for young Aboriginal people otherwise stigmatised. Felton *et al.* describe the value of artisan social networks in outer suburbs beyond mercenary concerns. In Wangaratta, Glen Innes and Broken Hill, creativity is less about civic 'buzz' and more a product of the hard slog and personal passion of key individuals. Mayes shows how a range of practices inform creativity rather than the other way around—creativity is 'something to do' to bring people together for reasons other than promoting the creative industries per se. But rather than deflate a sense of why the creative industries are important (because they are not so commercial), these articles suggest new ways in which creative industries research can be enlivened and made social, by not assuming a capitalist-orientated language of firms, growth, employment and export and instead valuing the communitarian purposes to which creativity can be put. From these social goals, highly visible local creative industries emerge. Indeed, perhaps there is something in the conduct of research in remote, rural and socio-economically disadvantaged places that brings into focus, much more sharply than in places of prosperity and cultural wealth, that it is ultimately people that together and through their individual and shared activities constitute what we call the creative economy.

NOTE

[1] Much has been written about the conflation of 'cultural' and 'creative' in recent research in geography, urban studies and economics. For an overview of the debate about terminology and its effect on research practice, see Gibson and Kong (2005), Markusen *et al.* (2008) and O'Connor (2009). In this special issue, we concentrate especially on 'creativity' because of the broader intellectual debate about what comes to be understood as 'creative' outside major cities. That is not to say that 'cultural industries' is a redundant phrase—indeed, in many respects it is preferable to 'creative industries' because 'culture' signifies links to non-economic relations and is central to discourses shaping conditions and patterns of work (see Gibson 2003 for further discussion).

REFERENCES

BARNES, K., WAITT, G., GILL, N. & GIBSON, C. (2006) 'Community and nostalgia in urban revitalisation: a critique of urban village and creative class strategies as remedies for social "problems" ', *Australian Geographer* 37, pp. 335–54.

BATHELT, H. & GRÄF, A. (2008) 'Internal and external dynamics of the Munich film and TV industry cluster, and limitations to future growth', *Environment and Planning A* 40, pp. 1944–65.

BATTERSBY, B. & EWING, R. (2005) 'International trade performance: the gravity of Australia's remoteness', Treasury working paper, Federal Government Department of Treasury, Canberra.

BELL, D. & JAYNE, M. (2006) 'Conceptualising small cities', in Bell, D. & Jayne, M. (eds) *Small cities: urban experience beyond the metropolis*, Routledge, London, pp. 1–18.

BELL, D. & JAYNE, M. (forthcoming) 'The creative countryside: policy and practice in the UK rural cultural economy', *Journal of Rural Studies*.

BOURDIEU, P. (1993) *The field of cultural production*, Polity Press, Cambridge.

BRENNAN-HORLEY, C. & GIBSON, C. (2009) 'Where is creativity in the city? Integrating qualitative and GIS methods', *Environment and Planning A* 41, pp. 2595–614.

BROWN, A., O'CONNOR, J. & COHEN, S. (2000) 'Local music policies within a global music industry: cultural quarters in Manchester and Sheffield', *Geoforum* 31, pp. 437–51.

CASELLAS, A. & PALLARES-BARBERA, M. (2009) 'Public-sector intervention in embodying the new economy in inner urban areas: the Barcelona experience', *Urban Studies* 46, pp. 1137–55.

CHRISTEN, K. (2008) 'Tracking properness: repackaging culture in a remote Australian town', *Cultural Anthropology* 21(3), pp. 416–46.

CHRISTOPHERSON, S. & STORPER, M. (1986) 'The city as studio; the world as back lot: the impact of vertical disintegration on the location of the motion picture industry', *Environment and Planning D: Society and Space* 4(3), pp. 305–20.

CONNELL, J. & GIBSON, C. (2003) *Sound tracks: popular music, identity and place*, Routledge, New York.

CONNELL, J. & GIBSON, C. (2009) 'Ambient Australia: music, meditation and tourist places', in Johansson, O. & Bell, T. (eds) *Sound, society and the geography of popular music*, Ashgate, Aldershot, pp. 67–88.

DAVIES, A. (2008) 'Declining youth in-migration in rural Western Australia: the role of perceptions of rural employment and lifestyle opportunities', *Geographical Research* 46(2), pp. 162–71.

EDENSOR, T., LESLIE, D., MILLINGTON, S. & RANTISI, N. (eds) (2009) *Spaces of vernacular creativity: rethinking the cultural economy*, Routledge, London.

EVERSOLE, R. (2005) 'Challenging the creative class: innovation, "creative regions" and community development', *Australasian Journal of Regional Studies* 11(3), pp. 351–60.

FLORIDA, R. (2002) *The rise of the creative class*, Basic Books, New York.

GABRIEL, M. (2002) 'Australia's regional youth exodus', *Journal of Rural Studies* 18, pp. 209–12.

GIBSON, C. (2002) 'Rural transformation and cultural industries: popular music on the New South Wales Far North Coast', *Australian Geographical Studies* 40(3), pp. 336–56.

GIBSON, C. (2003) 'Cultures at work: why "culture" matters in research on the "cultural" industries', *Social and Cultural Geography* 4(2), pp. 201–15.

GIBSON, C. & BRENNAN-HORLEY, C. (2006) 'Goodbye pram city: beyond inner/outer zone binaries in creative city research', *Urban Policy and Research* 24, pp. 455–71.

GIBSON, C. & CONNELL, J. (2004) 'Cultural industry production in remote places: Indigenous popular music in Australia', in Power, D. & Scott, A. (eds) *The cultural industries and the production of culture*, Routledge, London and New York, pp. 243–58.

GIBSON, C. & KLOCKER, N. (2004) 'Academic publishing as "creative" industry, and recent discourses of "creative economies": some critical reflections', *Area* 36(4), pp. 423–34.

GIBSON, C. & KLOCKER, N. (2005) 'The "cultural turn" in Australian regional economic development discourse: neoliberalising creativity?', *Geographical Research* 43, pp. 93–102.

GIBSON, C. & KONG, L. (2005) 'Cultural economy: a critical review', *Progress in Human Geography* 29(5), pp. 541–61.

HANSEN, H. & NIEDOMYSL, T. (2009) 'Migration of the creative class: evidence from Sweden', *Journal of Economic Geography* 9(2), pp. 191–206.

HIGGS, P.L. & CUNNINGHAM, S.D. (2007) *Australia's creative economy: mapping methodologies*, Queensland University of Technology, Brisbane.

JOHNSON, L. (2006) 'Valuing the arts: theorising and realising cultural capital in an Australian city', *Geographical Research* 44(3), pp. 296–309.

JOHNSON, L. (2009) *Cultural capitals: revaluing the arts, remaking urban spaces*, Ashgate, Aldershot.

KNEAFSEY, M. (2001) 'Rural cultural economy: tourism and social relations', *Annals of Tourism Research* 28, pp. 762–83.

LEADBEATER, C. & OAKLEY, K. (1999) *The independents: Britain's new cultural entrepreneurs*, Demos, London.

LOVERING, J. (2001) 'The coming regional crisis (and how to avoid it)', *Regional Studies* 35(4), pp. 349–54.

MACLEOD, G. (2002) 'From urban entrepreneurialism to a "revanchist city"? On the spatial injustices of Glasgow's renaissance', *Antipode* 34, pp. 602–24.

MARKUSEN, A. (2007) 'A consumption base theory of development: an application to the rural cultural economy', *Agricultural and Resource Economics Review* 36, pp. 9–23.

MARKUSEN, A., WASSALL, G.H., DeNATALE, D. & COHEN, R. (2008) 'Defining the creative economy: industry and occupational approaches', *Economic Development Quarterly* 22, pp. 24–45.

McGREGOR, A. & GIBSON, C. (2009) 'Musical work in a university town: the shifting spaces and practices of DJs in Dunedin', *Asia-Pacific Viewpoint* 50(3), pp. 277–88.

McLEAY, C. (1994) 'The "Dunedin Sound"—New Zealand rock and cultural geography', *Perfect Beat* 2, pp. 38–50.

MYERS, F. (2002) *Painting culture: the making of an Aboriginal high art*, Duke University Press, Durham, NC.

O'CONNOR, J. (2009) 'Creative industries: a new direction?', *International Journal of Cultural Policy*, 15(4), pp. 387–402.

OSBORNE, T. (2003) 'Against "creativity": a philistine rant', *Economy and Society* 32, pp. 507–25.

POWER, D. & JANNSON, J. (2008) 'Outside in: peripheral cultural industries and global markets', in Bærenholdt, J. & Granås, B. (eds) *Mobility and place: enacting Northern European peripheries*, Ashgate, Aldershot, pp. 167–79.

REIMER, S., PINCH, S. & SUNLEY, P. (2008) 'Design spaces: agglomeration and creativity in British design spaces', *Geografiska Annaler B: Human Geography* 90B, pp. 151–72.

SCOTT, A.J. (1984a) 'Industrial organization and the logic of intra-metropolitan location 3: a case study of the women's dress industry in the greater Los Angeles region', *Economic Geography* 60, pp. 3–27.

SCOTT, A.J. (1984b) 'Territorial reproduction and transformation in a local labor-market: the animated film workers of Los Angeles', *Environment and Planning D: Society and Space* 2, pp. 277–307.

SCOTT, A.J. (1999) 'The cultural economy: geography and the creative field', *Media, Culture and Society* 21, pp. 807–17.

SCOTT, A.J. (2000) *The Cultural economy of cities*, Sage, London.

SCOTT, A.J. (2006) 'Entrepreneurship, innovation and industrial development: geography and the creative field revisited', *Small Business Economics* 26, pp. 1–24.

SCOTT, A.J. (forthcoming) 'Cultural economy of landscape: pathways of development in the English Lake District', unpublished manuscript, University of California, Los Angeles.

SORENSEN, T. (2009) 'Creativity in rural development: an Australian response to Florida (or a view from the fringe)', *International Journal of Foresight and Innovation Policy* 5(1–3), pp. 24–43.

STRATTON, J. (2008) 'The difference of Perth music: a scene in cultural and historical context', *Continuum: Journal of Media & Cultural Studies* 22, pp. 613–22.

WAITT, G. & GIBSON, C. (2009) 'Creative small cities: rethinking the creative economy in place', *Urban Studies* 46(5&6), pp. 1223–46.

WATSON, A. (2008) 'Global music city: knowledge and geographical proximity in London's recorded music industry', *Area* 40, pp. 12–23.

YOUNG, C., DIEP, M. & DRABBLE, S. (2006) 'Living with difference? The "cosmopolitan city" and urban reimaging in Manchester, UK', *Urban Studies*, 43(10), pp. 1687–714.

Postcards from Somewhere: 'marginal' cultural production, creativity and community

ROBYN MAYES, *Curtin University, Perth, Australia*

ABSTRACT   *This paper focuses on a case study of local postcard production in a rural community in Western Australia. Drawing on in-depth interviews with key producers of these postcards, the analysis presented explores perceptions of and contexts for the emergence of this production, in turn examining the notion of 'creativity' articulated and privileged by this cultural work. Connections are identified between the making of postcards, the broader historical field of local cultural work and the construction of community. This, in turn, forms the basis for consideration of the role and relativity of 'marginality'.*

Introduction

In the summer heat of January 2007, after driving for several hours on seemingly endless country roads through sparsely inhabited bush and farmland, I arrived in the rural Shire of Ravensthorpe, Western Australia. This was to be the first of numerous enjoyable visits over the following two years as part of an ethnographic study of community change and place identity. First settled in 1868, the Shire of Ravensthorpe lies some 530 km south-east of the capital city of Perth (Shire of Ravensthorpe 2009). Agriculture, in particular broad-acre farming of grains and sheep, has consistently been a principal industry; farming continues by third- and fourth-generation pioneer families (Archer 2008; Southern Scribes 2000) and also by first- and second-generation settlers who arrived in the 1960s to take up new land farm blocks released, for example, under the Conditional Purchase Scheme (Williams 2009). Mining has been a part of the Shire from the discovery of gold in the late 1890s, through to the more recent and ultimately brief operational presence of a large-scale BHP Billiton nickel mine which opened in May 2008 and closed in January 2009 (Shire of Ravensthorpe 2009). Prior to the rapid increase in population and profound demographic change occurring as a result of the accommodation of some 350 mine staff and families, the Shire's residents numbered 1387 in 1996, 1466 in 2001 and 1951 in 2006 (ABS 2007). This small population is dispersed among the communities of Hopetoun (a retirement coastal settlement and, briefly, home to the majority of residential mine staff);

Ravensthorpe (seat of local government and service centre for surrounding agricultural industries); Jerdacuttup (a small community farming land released in the 1960s), Munglinup (a farming community which is also part of Esperance Shire) and Fitzgerald (also a farming community and which lost its school recently owing to a lack of students). Over two-thirds of the Shire's 13 000 km^2 remains as natural bush land and national parks, including the eastern half of the Fitzgerald Biosphere Reserve (Shire of Ravensthorpe 2006).

On this first visit to this Shire I was fascinated, and surprised, to discover the Ravensthorpe Visitor Centre and Museum (open on a Sunday) overflowing with locally produced books, wrapping paper, screen-printed tea towels, jewellery, and stands of postcards, to name just a small selection of goods. In this and subsequent visits I learned of the annual Wildflower Show and local herbarium. I also learned that the Centenary Mural Sculpture on the main street was locally designed and crafted and that the home page image on the Shire website is a local landscape painted by a resident artist. I admired the community-developed Hopetoun entry sign and artwork along with the public wall murals painted by local community groups. I was introduced to numerous groups such as the Southern Scribes, the Jerdacuttup Players, and the Boot Scooting Club. During my time in the area I have had the pleasure of attending plays, book launches, the 25th Annual Wildflower Show, festivals, community concerts, and have conducted some of my interviews in the Jerdacuttup Community Hall surrounded by an impressive array of theatre arts equipment. Through friends in the area and the reading of the local newspaper I am aware of numerous events that I wasn't fortunate enough to attend: week-long writer-in-residence programs, a performance of a locally authored opera, numerous workshops and, more recently, the opening of the Ravensthorpe District Art Group Gallery Dunnart Studio Gallery.[1]

My surprise at finding this rich and vibrant cultural life, this abundance and range of cultural enterprise and production, is not only a somewhat embarrassing manifestation of my (urban) ignorance if not arrogance; it can also be read as a symptom of culturally widespread assumptions which privilege urban Australia as a normative site of and place for creativity and cultural industry. This paper explores cultural work in Ravensthorpe in relation to urban-centric discourses around the cultural industries elaborated in the introduction to this volume. It does so through focusing on the local production of postcards, all of which feature images from the area, principally land and seascapes, flora, and town buildings (see Plates 1–6). These postcards are available exclusively from town vendors such as the Ravensthorpe Post Office, Hopetoun Telecenter, and Ravensthorpe Visitor Centre. With a 30-year history and involving 20 residents (13 of whom are female) across this period, this production includes the work of individuals and voluntary community committees such as Ravensthorpe Hopetoun Area Promotions (RHAPS), and informal art groups. As I have argued elsewhere (Mayes 2009), this local production of postcards plays an important role in the ongoing construction and circulation of place identity and in the representation and embodiment of rurality more broadly. To make postcards is also to participate in and engage with a substantial and highly conventionalised cultural industry and communicative practice with a significant role in the (potentially mass) production and circulation of (popular) meaning (Kennedy 2005; Pritchard & Morgan 2005; Sigel 2000). Aimed squarely, though not exclusively, at an external market/ audience and partly a response to marginalisation by the commercial postcard

PLATE 1. Postcard locally designed and printed by Jan K. Enterprises.

industry—none of the commercial firms offer postcards carrying images of the Ravensthorpe region—this postcard production is suggestive of the multi-scalar spaces and practices (and possible specificities) of rural cultural industries and creativity. The analysis offered here pays particular attention to the forms and contexts of creativity encoded in and enabling this production, together with the role of geographical distance and cultural isolation in its emergence. The paper begins with a brief description of the empirical data collection and then situates the study within broad understandings of 'cultural industry' and 'creativity'. Thereafter, the postcard producers' perceptions of, and local contexts for, this production are examined in light of understandings of, and linkages between, creativity and community. The paper concludes by mapping some of the ways in which this 'marginal' practice engages with dominant and alternative models of cultural industry.

PLATE 2. Postcard locally designed and printed by Jan K. Enterprises.

PLATE 3. Postcard locally designed and printed by Shirl and Graeme Sutherland.

Case study and context

Though drawing on the wider ethnographic study mentioned previously, primarily in order to elucidate broader social, cultural and historical contexts, the discussion is grounded in five semi-structured interviews conducted as part of this broader project. These five interviewees were selected on the basis of substantial involvement in the production of local postcards. All interviewees are female and four are long-term local residents living in the area upwards of 40 years, while one

PLATE 4. Postcard locally designed and printed by Jan K. Enterprises.

**The Old
Hopetoun Jetty**

This jetty served the Hopetoun, Ravensthorpe district from 1902 to 1983. The Phillip's River gold rush at the turn of the century brought ships to tie up at the jetty's solid jarrah timbers; all freight came by sea. Wool, gold and copper ore were shipped out. In 1908 a railway was constructed linking the two towns and the line was added to the jetty. By 1909 the population of Ravensthorpe had reached 3,000 and there was a daily train service. 83,941 ounces of gold had passed over the deck of the jetty by 1918. The First World War, end of large scale mining and the depression finally led to the closure of the railway and the last commercial ship called at Hopetoun in 1937. Declared unsafe the jetty was burnt in 1983.

PLATE 5. Postcard locally designed by Richenda Goldfinche, 'Old Hopetoun Jetty'. Pastel on coloured paper.

interviewee had been living in the community close to 2 years. One individual producer of an extensive range of postcards, and the only male with substantial involvement, could not be contacted for interview. Though this sample is small it nevertheless captures a sizeable percentage of local producers, and spans not only individual and collaborative participation but also past and present production. Interviews were semi-structured around reasons for and benefits of becoming involved in making postcards, the processes of production including the ways in which images were selected, and their market success in terms of sales and most popular images. Interviews were recorded and transcribed.

The terms 'cultural' and 'creative', as is widely argued in the literature, are often uncritically conflated and at risk of becoming so capacious as to become meaningless (Galloway & Dunlop 2007; Warde 2002). As such they require some clarification in order to delineate the conceptual boundaries of the ensuing analysis. Precise definition of cultural industries, however, is far from easy (Drake 2003). Inclusion of a specific activity involves a complex qualitative assessment

PLATE 6. Postcard locally designed by Ravensthorpe Hopetoun Area Promotions, photographs by Helen Taylor.

necessarily informed by social and spatial contexts (Banks 2007). Notwithstanding this and further difficulties around determining relative proportions of cultural and functional worth, the category 'cultural industries', as a distinctive subset of the broader creative industries, is defined not only as privileging symbolic meaning (over 'use value'), and thus a distinctive cultural creativity, but in addition as employing large-scale or mass production methods (Banks 2007; Galloway & Dunlop 2007). Much of the Shire of Ravensthorpe's cultural production described above sits squarely in the 'creative arts' sector: together with the use of semi-industrial and small-scale production methods, this positions the cultural work undertaken in Ravensthorpe as 'peripheral' to the cultural industries (Galloway & Dunlop 2007). Similarly, 'creativity' is seemingly resistant to consensual definition, though the literature convincingly emphasises its collective and social roots (Bilton & Leary 2002; Drake 2003) and the necessary role of social capital (Scott 1999).

Making postcards in Ravensthorpe

Interviewees were not asked directly about 'creativity' and none of them used the term in any context. Rather, interviewees noted personal characteristics such as having 'a mind that just keeps ticking over' so that the postcard work is 'just a little hobby to keep me out of mischief'. In the words of another interviewee: 'it's an interest'. Importantly, this interest is elaborated upon as facilitating interactions with other groups; one interviewee gave the example of having taken a photograph of 'an unusual flower' which then led to discussions with herbarium enthusiasts in an attempt to identify the species and understand why this particular specimen did not exhibit typical petal structures. Yet another interviewee became involved by virtue of committee membership:

> I found myself on the committee, not that I offered, but it just sort of happened. They just left it to me. And I just sort of took all sorts of photographs and then we just put them in and said well we'll use this one and this one and this one.

In commenting on how she came to be involved in making postcards, a further participant noted that 'we had a strong art group'. Romantic notions of the creative self, in which creativity arises naturally from essentialised and unique individual qualities (Banks 2007) are noticeably absent in these fundamentally self-effacing narratives. At the same time, the postcard work appears to be 'marginal'. It is just one local cultural enterprise among others. This stands in contrast to the creative focus on specialised forms described in the literature on (competitive) clusters around, for example, film making or jewellery production. The postcard work is further marginal in relation to the rest of the producer's life, as in the designation 'hobby'. Similarly, self-expression and conscious meaning-making, and to a lesser extent the communication of ideas—hallmarks of culturally dominant under-standings of individual creativity—were underplayed by interviewees. Instead, emphasis was afforded to a desire to 'share' and, in the case of RHAPS production, to promote the area and fundraise for this purpose. In the case of the art collective, attention was concurrently on promotion of local artists, consistent with postcards as advertising and promotional artefact (Kohn 2003), and on redressing the above-mentioned local lack of postcards.

At some point in the discussion around how she became involved in making postcards, each interviewee made (often extensive) reference to the broader community, and to other cultural activities and products. They mentioned seemingly unrelated affiliations with, for example, the Southern Scribes or the Wildflower Show together with assertions about broader community involvement encapsulated in the following assertion: 'I'm pretty much a community person.' This production occurs not only in a broader field of local cultural work and community networks (Mayes 2009) but is also situated in a particular 'type' and history of community. The narratives offered by two producers with long-term residencies in the Shire draw attention to the intertwined emergence of community and cultural work. This history begins with the arrival of new land farmers in the 1960s in Fitzgerald and Jerdacuttup. The method and allocation of this new land ensured that settlement occurred over a short period so that, in the words of one interviewee, in each area it was 'all basically new people'. In Fitzgerald 'a lot of people were from the Eastern States'. Those taking up new land in the Jerdacuttup area are described as coming

> from all over the world really, you know. From Africa, from England, from America, from Switzerland, from Eastern States etc. and people from different walks of life—some in agriculture, some not, but all ending up with land. And it was quite eclectic; you know it was really quite exciting. Because we had a great mix. And we found lots of things started happening because of this.

These newcomers, implicitly, are not 'just' farmers, just as 'newness' and 'difference' are important characteristics not only in 'starting a community' but also in starting cultural production in order to build community. As was explained:

> we started a community with a whole group of us, we were all new land farmers together, nobody had any money ... First of all, the community got together and raised money ... so that was the first part, was that we raised money and put money in to build a hall, school come church, come everything, but it was mainly for children. We had dances, and pictures, and very much to the fore of helping raise money for things and community-minded.

An important aspect of starting these communities was not only lack of infrastructure—there was literally nothing—but also distance from other settled areas and amenities. The task for the newcomers, according to these interviewees, and as numerous others in the broader ethnography proudly confirm, was to provide themselves with both infrastructure and a sense of community through shared work and social events. Fundraising through local cultural work was an ideal solution. As an example, the Jerdacuttup Players theatre group emerged from a Music Hall performance staged in 1972 as a fundraiser for a power plant for the local hall (Hatter 2009). Since then the group has produced 46 plays, including at least one locally authored piece between 1972 and 2006. These were performed in Jerdacuttup, Hopetoun and Ravensthorpe, occasionally in Munglinup, Lake King and, further afield, in Esperance. In this manner, the Players, together with other voluntary local community groups such as the Ladies Catering Group, funded improvements to the hall including 'dressing rooms, toilets, a supper room, tables, chairs, costume cupboards and stoves' (Hatter 2009, p. 341), not to mention the specialised equipment and resources needed to maintain a viable drama group. These events not only require widespread community involvement in a range of creative, supportive and organisational roles, but also occur in, and reproduce, a sympathetic environment which in turn enabled connections with the other communities:

> there used to be that comradeship-come-support feeling you know? If a community put on an event, you supported it 'cause you knew how much time and effort it took to put on an event. You always supported an event that was put on.

The community postcard production as a fundraising strategy (Mayes 2009) is part of this tradition of cultural work in the service of community needs. Presented by interviewees as marginal, though important, to their individual lives, cultural work and products more broadly are concurrently understood as central to 'community'.

Community group and individual postcard production occurs also in what can be termed a 'creative' environment. As noted above, one range of postcards came into existence through the proximity of other artists. Another interviewee pointed out that, 'Well we've got a lot of talented people in this place', suggesting that the presence of like-minded or talented others in some proximity is an enabling factor. Attention, in addition, was also drawn to a sense of place-based differences within the Shire. Jerdacuttup, in particular, is perceived both internally and externally as a locus of creative, if not bohemian, difference:

> Ravensthorpe always thought we were immoral and newcomers, you know were rather shocked by our goings on.

A sense of heightened community investment in cultural production informs this understanding of difference:

> We have an Arts Council which has always been Jerdacuttup based; it's called the Ravensthorpe Arts Council but it was started by Jerdacuttup people and is still run by Jerdacuttup people and we've put on shows here and for years, and years, and years and years.[2]

Each of the four long-term local participants had extended, close involvements with this 'new land' community.

Jerdacuttup, as a close-knit, 'artistic' and politically engaged community with a strong cultural-arts tradition, is confirmed in the broader ethnography as a perception, widely held in the Ravensthorp area. This difference is seen as valuable, and also as an irritation, as demonstrated in the following comments from a variety of residents from the other communities in the Shire:

> Their plays are absolutely brilliant. Everybody raves about them. They've really got some really talented local farmers ... local people there.

> When Jerdacuttup was a very new area they had a very strong art group down there. And they used to have film evenings and drama groups and whatever ... that generation has continued to foster that in their community and they still hold plays and they even had an open air concert last year. So they've really ... they have retained their identity through that group. It's quite a great little group and I mean that's not the only thing, but there's quite a few artistic people down there that have really worked to maintain their identity.

Jerdacuttup was always an elite arty-farty group. That's the way to describe them, the elite arty-farty group. And if there was going to be something ... how do I say it? Perhaps if they were going to go and save an orchard, it'd come from there.

Whether valued and appreciated or disparaged, Jerdacuttup is locally experienced and perceived as a hub of creative endeavour in a process in which creativity is no longer marginal.

Creativity and communities on the margins

Interviewees' narratives around their involvement in the production of postcards, whether as individuals or as part of a committee, support an understanding of creativity as quotidian activity in the service of solving daily problems (such as lack of local postcards, the need to keep busy, fundraising) with an emphasis not so much on novelty but rather on 'usefulness' (e.g. Bilton & Leary 2002). In this understanding, creativity is a 'multidimensional' and collective, if not collaborative, process arising from combinations of factors and relationships. The postcard production in Ravensthorpe Shire foregrounds a variety of layered motivations and relationships, thus suggesting the 'diversity and contingency of forms of creative entrepreneurial behaviour' (Drake 2003 citing MacKinnon et al. 2000). In addition, this local practice indicates that the 'kind' of creativity and the conditions of emergence are closely interlinked. The connections to broader networks, projects and histories of community woven into the narratives presented here emphasise isolation or more specifically 'lack' as a positive stimulus when combined with a desire/need to create community; this isolation or lack is not individual but is rather geographic community isolation. Consequently, the Ravensthorpe Shire postcard work and broader context highlights the 'sociality of cultural industries' (Kong 2005; Scott 1999). This incorporates social bases, premised on the understanding that cultural production and consumption is embedded in social systems and relationships, along with social roles such as 'contributing to the development of cities as cultural and social entities', that is 'as places where people meet, talk, share ideas and desires, and where identities are formed' (Kong 2005 citing Bianchini 1993). The postcard production in Ravensthorpe Shire draws on a considerable store of accumulated social and symbolic capital. This regionally unique cultural work has its roots in substantial (and historical) local interactions both within and across communities, and a broader communal sense of, and pride in, creative endeavour. The limits of this endeavour, it is reasonable to surmise, are reproduced and negotiated in shared highly subjective perceptions of what is 'fun', will serve the community best, will generate the most funds and opportunities for skills development, and make best use of existing local skills and resources. Place-based networks such as this tend to be associated with cities presented as logical sites of (larger-scale, economically potent) aggregation (Kong 2005).

The overarching objective of establishing a sense of community, as historical context for local cultural production including but not limited to postcard production, in turn emphasises collaborative creativity around cultural products for local consumption *and* participation. This local work is also suggestive of creativity as a means to enhance interaction rather than ('just') interaction as a means to enhance creativity, as privileged in much of the literature (see Drake 2003).

Creativity, as defined above and in the eyes of the postcard producers interviewed, is in the service of 'community', so that creativity has a clear social role in the production of a sense of community and the construction of both physical places and social spaces in which to meet, share ideas and negotiate individual and community identities. This is further evidenced by the wider local connections articulated between 'artistic' and 'community'—in particular community identity. Cultural industries offer opportunities for the negotiation and transformation of rural identities (Gibson 2002). The construction of Jerdacuttup as a discrete community with distinctive social capital through creative practice—politically important for a small farming community which, in the words of one interviewee, has 'always been a small community' and 'always had to battle for things'—highlights creativity as both a social and political process (Banks 2007). It is not only community recognition, if not standing, that is achieved; individual members also are fashioned as 'talented', for example. It is important to keep in mind that this effacement of individual motivations and emphasis on community in these interviews may be informed by a gendered sense of self. The absence of the term 'creative' in interviewee descriptions of their work or selves may similarly have its roots in gendered experience. At the same time, however, 'artistic' is used to describe both the local cultural activity and individual personal characteristics, thus reinstating the Romantic notion of the creative individual. Worthy of further exploration, connections are evident between this 'artistic' practice and generalised critical politics as seen in the community perceptions of Jerdacuttup.

Postcard producers have either personal histories of involvement in numerous and various community activities or are able to activate the networks and, equally importantly, the ethos this has established. The newcomer, too, has been able to find a 'space' in this environment and is able to utilise local social capital. Individual enterprise thus enacts a fluid interaction between collective creativity and networks as a fertile environment for individual production (see Drake 2003). There is no evident tension between the volunteer community producers and individual producers either in terms of access to market share or sense of validity of product as 'authentically' local.

The postcard producers, particularly in light of the range of other cultural work they undertake, and together with the broader group of local people engaged in cultural production of various forms, can be understood as 'a community of creative workers' (Drake 2003)—an important feature of 'locality' in the context of cultural industries. These participants are directly engaged in the production of commodities in which symbolic meaning is privileged, and do so collectively, and potentially as a 'creative class'. This status indicates possession of desirable/sought after skills available as a 'talent pool' (Bontje & Musterd 2009); implicit in this are attendant exclusions and uneven power relations. Indeed, the foregrounding of (enabling and empowered) community is not to suggest that the community called into being in this practice is inclusive, apolitical or utopian. There is also the matter of who can contribute (aside from perceptions or evidence of talent); ability to participate in independent (and here largely unpaid) cultural work has much to do with 'lifecycle' stages (McRobbie 2002). The founding members of these new land communities were predominantly young (middle-class) couples either with young families or on the verge of starting families. The postcard work is currently the domain of a retirement-age cohort and, as indicated at the outset, is dominated by women. Farming is also an industry that lends itself to community development: first, in this area the majority of farmers were landowners developing their own businesses with attendant autonomy; second, land management does not stop at the fence and requires collaboration with neighbours; third, this group was all similarly cash-poor; and, fourth, farming is a cyclical business with down-time.

Conclusion: engaging markets and practice-led cultural industries

As has been demonstrated convincingly by numerous scholars, there are 'inherent tensions in the economic and social roles of cultural activities' (Kong 2005, p. 65). Broadly, the 'creative economy', though premised as a means to 'overcome social and economic inequalities and effect future economic growth' (Banks 2007, p. 71), has been shown to privilege 'economic growth' while neglecting those who cannot participate and, at the same time, effecting active exclusions (Atkinson & Easthope 2009). Importantly, as Galloway and Dunlop (2007, p. 29) argue, positioning 'the cultural sector as part of the wider creative economy simply subsumes it within an economic agenda', thus occluding the public benefits provided by culture. More specifically, making money, as overarching objective in the creative economy, 'structures how creativity is defined, developed and employed' (Banks 2007, p. 73). The production of postcards in the Shire of Ravensthorpe, whether undertaken by individuals or by community groups, does not have money making as its central objective (Mayes 2009). Even so, the postcard production is, to some extent, about finding a place in the market for postcards, just as it requires for its existence, and forms links to, (global) tourism markets and capital flows. While postcard producers are focused on meeting very local needs, production is not entirely contained by this context; one individual producer has ambitions to extend her production to represent and sell from other regions, another producer had to change production methods in order to comply with commercial standards of production and another producer felt the need to note that 'it's done like a proper postcard'. In this way the market informs this local practice which in turn does not/cannot operate outside the market.

Meanwhile, this production has been shown to reject normative capitalist imperatives to focus on maximising profits and to strive for wealth accumulation, for example in producers' marginal interest in, if not disregard for, financial gain (in part encoded in choice of product). Rather, this local production emphasises internal (participatory) rewards as associated with 'practice led economies' (Banks 2007). Though unlikely to qualify as 'alternative' production, the practice-led emphases in this postcard work offer at least a glimmer of 'utopian promise' which Banks (2007) insists imbues creative cultural production, despite convincing arguments demonstrating the pervasive (if not total) capitalistic appropriation of the culture industry. The potential for cultural criticism or subversion is similarly difficult to assess, though this rural cultural work and creativity may well provide a source of, or creative field for, stimulating not only 'new' but also alternative cultural products.

This case study of postcard production points to the importance of recognising and understanding rural individual and community cultural work as a wide-reaching and multi-faceted component of rural and community life, and also as providing insights into progressive and transformative potentials in the creative industries. Within the broader normative conceptualisation of the cultural industries, the postcard production undertaken in Ravensthorpe is marginalised on several counts: it is non-urban, non-industrial, non-market oriented. At the same time, this activity occurs in conditions similar to those identified as underpinning urban or clustered cultural industry: networks, stimulus, and community traditions. This sort of activity may well be taking place in urban communities, but in this rural community it is not only highly visible but also *central* to the material and symbolic realities of community life. Rural places and communities have much to teach us about creativity and cultural production, particularly as occurs when (substantially) decoupled from 'economy', along with the means by which the contemporary reduction of the cultural industries to a capitalist economy might be resisted.

Acknowledgements

I would like to extend a warm 'thank you' to the 'postcard producers of Ravensthorpe Shire' for sharing their experiences with me and commenting on the draft paper. Sincere thanks also to Chris Gibson for the invitation to contribute to this issue. Much of this research was conducted during an independent Fellowship funded by the Alcoa Foundation's Conservation and Sustainability Program at the Alcoa Research Centre for Stronger Communities at Curtin University of Technology.

NOTES

[1]   This list is far from complete and is not intended to suggest that the noted items are more 'cultural' or in any other way worthier than those not mentioned.
[2]   This subjective view is not offered as historically accurate.

REFERENCES

ARCHER, A.W. (2008) *Ravensthorpe then and now* Third revised edition, Ravensthorpe Historical Society, Ravensthorpe.

ATKINSON, R. & EASTHOPE, H. (2009) 'The consequences of the creative class: the pursuit of creativity strategies in Australia's cities', *International Journal of Urban and Regional Research* 33(1), pp. 64–79.

AUSTRALIAN BUREAU OF STATISTICS (ABS) (2007) *2006 census tables: Ravensthorpe (S) LGA* (accessed August 2007).

BANKS, M. (2007) *The politics of cultural work*, Palgrave Macmillan, Houndmills, Basingstoke.

BILTON, C. & LEARY, R. (2002) 'What can managers do for creativity? Brokering creativity in the creative industries', *International Journal of Cultural Policy* 8(1), pp. 49–64.

BONTJE, M. & MUSTERD, S. (2009) 'Creative industries, creative class and competitiveness: expert opinions critically appraised', *Geoforum* 40(5), pp. 843–52.

DRAKE, G. (2003) '"This place gives me space": place and creativity in the creative industries', *Geoforum* 34, pp. 511–24.

GALLOWAY, S. & DUNLOP, S. (2007) 'A critique of definitions of the cultural and creative industries in public policy', *International Journal of Cultural Policy* 13(1), pp. 17–31.

GIBSON, C. (2002) 'Rural transformation and cultural industries: popular music on the New South Wales far north coast', *Australian Geographical Studies* 40(3), pp. 337–56.

HATTER, M. (2009) 'Jerdacuttup Players', in Williams, A. (ed.) *Where the boodie rats used to dig: a scrapbook history of Jerdacuttup pioneer farming community*, Jerdacuttup Community Association, Jerdacuttup, WA, pp. 340–43.

KENNEDY, C. (2005) 'Just perfect! The pragmatics of valuation in holiday postcards', in Jaworski, A. & Pritchard, A. (eds) *Discourse, communication and tourism*, Chanel View, Clevedon, pp. 223–46.

KOHN, A. (2003) 'Let's put our (post) cards on the table', *Visual Communication* 2, pp. 265–84.

KONG, L. (2005) 'The sociality of cultural industries', *International Journal of Cultural Policy* 11(1), pp. 61–76.

MAYES, R. (2009) 'Doing cultural work: local postcard production and place identity in a rural shire', *Journal of Rural Studies* (in press).

McROBBIE, A. (2002) 'From Holloway to Hollywood: happiness at work in the new cultural economy?', in du Gay, P. & Pryke, M. (eds) *Cultural economy: cultural analysis and commercial life*, Sage, London, pp. 97–114.

PRITCHARD, A. & MORGAN, N. (2005) 'Representations of "ethnographic knowledge"; early comic postcards of Wales', in Jaworski, A. & Pritchard, A. (eds) *Discourse, communication and tourism*, Chanel View, Clevedon, pp. 53–75.

SCOTT, A. (1999) 'The cultural economy: geography and the creative field', *Media, Culture & Society* 21, pp. 807–17.

SHIRE OF RAVENSTHORPE (2006) *Community information: a few facts*, brochure available locally and from the Shire webpage: http://www.ravensthorpe.wa.gov.au/ (accessed 26 October 2006).

SHIRE OF RAVENSTHORPE (2009) 'Mining', available from: http://www.ravensthorpe.wa.gov.au/mining (accessed 20 January 2010).

SIGEL, L. (2000) 'Filth in wrong people's hands: postcards and the expansion of pornography in Britain and the Atlantic World, 1880–1914', *Journal of Social History* 33(4), pp. 859–85.

SOUTHERN SCRIBES (2000) *And the dingoes howled: stories from the bush*, Southern Scribes, Ravensthorpe.

WARDE, A. (2002) 'Production, consumption and "cultural economy"', in du Gay, P. & Pryke, M. (eds) *Cultural economy: cultural analysis and commercial life*, Sage, London, pp. 185 200.

WILLIAMS, A. (2009) *Where the boodie rats used to dig: a scrapbook history of Jerdacuttup pioneer farming community*, Jerdacuttup Community Association, Jerdacuttup, WA.

Creativity without Borders? Rethinking remoteness and proximity

CHRIS GIBSON, SUSAN LUCKMAN & JULIE WILLOUGHBY-SMITH, *University of Wollongong, New South Wales, Australia; University of South Australia, Adelaide, Australia; University of South Australia, Adelaide, Australia*

ABSTRACT *This article examines remoteness and proximity as geographical conditions and metaphors. It stems from a large government-funded research project which sought to examine the extent and uniqueness of the creative industries in Darwin—a small but important city in Australia's tropical Top End region, and government and administration capital of the sparsely populated Northern Territory. In talking to creative artists from diverse fields about their work and inspiration, it became clear that geographical positionality was a key framing device through which people understood themselves and their relationships with others. Remoteness and proximity were tangible in the sense of physical distances (Darwin is remote from southern States, and yet proximate to Asia and Aboriginal country). But Darwin's location was also perceived and imagined, in cultural texts, in creative workers' discussions of Darwin in relation to the outside world, and in their sense of the aesthetic qualities of the city's creative output (particularly shaped by multicultural and Aboriginal influences). We develop our analysis from 98 interviews with creative workers and postal surveys returned by 13 festival organisers in Darwin. Qualities of distance, proximity, isolation and connection materially shape a political economy of creative industry production, and infuse how creative workers view their activities within networks of trade, exchange and mobility.*

Darwin's image, since its beginnings, has been that of a government town, a place where exiles from the south fortified themselves with liquor against the loneliness and the heat, deserving of the jibe, attributed to Xavier Herbert, that its only exports were empty bottles and full public servants. (Powell 1988, p. 227)

Q: Where do you feel Darwin's strengths lie as a creative city?
A: I think its unique environmental and cultural make up, its proximity to South East Asia, its relative isolation from the rest of Australia. It is relatively intact in terms of culture and environment and I think it just

offers something that is pretty special and unique to the rest of Australia and the world. (Art gallery curator, interview, July 2007)

Introduction

Darwin is a frontier city. Fifteen hours' drive to the nearest substantial town (Alice Springs, with only 25 000 people) and 3000 km to the nearest State capital city (Adelaide), it was established by British colonisers on land taken from the Larrakia people, as a strategic northern outpost in an earlier geopolitical era when the 'yellow peril' and fear of northern invasion reigned (Jull 1991). Much has changed, not least from the multiple demolitions of war and cyclone. Darwin now trumpets itself as a cosmopolitan, if small (population 70 000), city with a multicultural population, modern infrastructure, busy international airport and substantial international tourism (Lea *et al.* 2009). Yet Darwin retains a palpable sense of solitude. Flights from other capital cities take several hours; in driving out of the city one quickly enters a brutal tropical savannah landscape: wide flooded coastal plains dominated by melaleuca forests, swampy tidal mangrove forests, and spiky pandanus. Its harsh climate of extreme heat and monsoonal storms has forced architectural adaptation and shaped the annual rhythms of economic and cultural activity. Darwin is an outpost of Australian federalism, designed principally around defence forces' needs, the delivery of government services, and the mining and pastoral industries. A northern capital retaining its strategic military, political and resource importance, in so many ways Darwin is tangibly—and metaphorically—remote.

What might this unusual setting mean for industries other than defence, government administration and mining—such as creative industries—that are common to cities everywhere, including Darwin, but which have not until now been critical to Darwin's strategic geopolitical *raison d'être*? This article stems from a research project where we sought to answer this question. Funded as an Australian Research Council Linkage Project (LP0667445; 2005–2009), the research project's three stated aims were to determine the nature, extent and change over time of the creative industries[1] in Darwin; to interrogate the applicability of national and international creative industry policy frameworks to Darwin; and to identify opportunities for local transformation in creative industries. Darwin's remote location, small size and relations between Aboriginal and non-Aboriginal groups were envisaged as deep challenges to existing orthodoxies in creative industries theory and policy making (Lea *et al.* 2009; Luckman *et al.* 2009). Research was needed to understand the challenges of remoteness for creative industries, when population mass and proximity to other major centres were absent, and in a complicated (post)colonial setting such as in Darwin.

For Darwin, remoteness presents particular challenges for creative workers—being a long way from key centres and scenes in Melbourne, Sydney and further afield. Staying in touch with key gatekeepers is difficult, as is maintaining visibility in these larger markets, and there is always pressure on talented up-and-comers to move to larger centres (see Bennett, this issue). And yet Darwin has its own forms of geographical proximity that favourably influence its cultural economy of creative industry production and distribution. While Darwin might seem remote to southern States, it is proximate to Asia and for over a century has had prominent Asian communities in its local demographic profile (with longer-established

Chinese communities now complemented by Philippine, Timorese and Indonesian populations). At the same time, Darwin is an Indigenous capital of the north, within a network of scattered, tiny Aboriginal settlements throughout Arnhem Land. Darwin is a centre for Aboriginal broadcasting, and for Indigenous visual artists, musicians, artefact producers and dancers. We sought to discover whether the flip-side of remoteness was a local cultural distinctiveness—a proximity rather than isolation, when viewed from 'inside' Darwin—borne of unusual combinations of cultures on the northern colonial frontier.

Also informing our project was a commitment to better understand the everyday experiences of workers in the creative industries. This commitment stemmed from ongoing critiques of creative city policy (which point to the tendency to over-glamorise creative work) and from research revealing how creative work is precarious and manifest differently across space (Gibson 2003; Gill & Pratt 2008; Ross 2008; Reimer 2009). Our focus on everyday experiences of creative workers required a 'grassroots' approach: creative workers literally produce the creative city through their everyday practices, working lives, movements, and imaginations of their city in wider cultural and economic flows (see Brennan-Horley, this issue). While in the national geopolitical imagination Darwin might be remoteness personified, those actually living and working there manage and contest that reality on a daily basis. We focus here on the manner in which creative workers grapple with remoteness and proximity and the difficulties these geographical conditions present. But we also seek to move beyond this, to discuss how distance is disavowed and alternative opportunities sought—how the constraints of isolation are tested, and its pleasures realised.

Method

After a pilot interview (with 14 people) in April 2007, 84 interviews were conducted over 11 months, with workers in a range of creative occupations—including activities normally associated with creative industries such as music, visual art and design, as well as vernacular creative occupations as contrasting as dance, tattooing and whip-making. Sampling methods included: web searches and listings in the Darwin Yellow Pages; recommendations from colleagues; snowbal-ling from one interviewee to others; and eventually receiving expressions of interest from potential interviewees (once word had spread about the project). Two interview schedules were used: one for those whose primary work was a creative activity and another for those whose primary source of income was not in the creative industries (see Luckman et al. 2009). Many of the questions were overtly geographical in nature, prompting participants to discuss where they lived, worked, networked, sought inspiration and saw creative activity concentrated in Darwin. Interviewees were asked to draw on a paper map of Darwin as they vocalised their responses to these questions. In-depth analysis of this embedded mapping component of our interviews is not a feature of this article (but see Brennan-Horley, this issue, and Brennan-Horley & Gibson 2009). Of relevance here, though, these questions sharpened interviewees' general geographical literacy across the whole interview. It is worth noting that the themes of this article—remoteness, proximity, isolation—were consciously discussed by interviewees possibly because we encouraged them to talk in geographical terms, using a map as visual prompt.

Beyond mapping questions, interviewees were asked about their experience of Darwin's creative industries and factors enabling and/or impeding Darwin as a creative city. Questions included:

- What is your involvement in Darwin's creative industries?
- Where do you feel Darwin's strengths lie as a creative city?
- What makes it easy to go about your business in Darwin?
- What impediments to creativity do you perceive in Darwin?
- What resources do you use as a creative worker?

These questions resonated in a context where policy development for creative industries has been intermittent and modest: the city has tended to favour mining and tourism as its main economic development concerns, with 'the arts' (rather than 'creative industries') receiving some government support as necessary civic cultural activities, part of the city's broader agenda to be a 'northern capital' rather than an economic development concern.

The final combined sample had the following characteristics: 89 per cent had their creative activity as their main occupation; slightly over 50 per cent were women; most were in their 30s or 40s (60 interviewees were aged between 30 and 50, with only nine under 30); interviewees were from 29 different ethnic backgrounds (including 10 people from several different tribal Aboriginal backgrounds) and were involved in 46 different creative activities (see Table 1). On balance, the sample was slightly older than was expected (probably a function of the predominance of full-time creative workers rather than part-timers or fledging young artists) and was a little disappointing for its proportional inclusion of Indigenous creative workers, although Aboriginal people were a notable presence in the sample. Statistical analysis of Darwin's creative workforce undertaken separately for the project revealed that Aboriginal people are numerically under-represented in the creative industries in Darwin compared to Darwin's total population (Gibson & Brennan-Horley 2007). That city-wide under-representation was reflected in our interview sample.

TABLE 1. Types of creative workers interviewed, 2007–08 (+ number of interviews)

Actor (5)	Dancer (2)	Multimedia designer (3)
Actor/director (1)	Designer (1)	Musician (7)
Advertiser (1)	Director (1)	Performing artist (4)
Architect (7)	Educator (4)	Photographer (6)
Artist (5)	Entertainment reporter (1)	Playwright (1)
Artistic director (5)	Festival and event manager (5)	Printmaker (1)
Arts administrator (22)	Filmmaker (2)	Promoter (3)
Arts retailer (2)	Goldsmith/jeweller (2)	Publishing (3)
Author (3)	Graphic designer (1)	Retail (2)
Building designer (1)	Horticulturalist and landscape	Sound artist (1)
Building surveyor (1)	consultant (1)	Sound engineer (3)
Choreographer (3)	Leatherworker (1)	Tattooist (1)
Cinematographer (2)	Manager (2)	Venue manager (1)
Clothing, bags and jewellery	Manufacturing and sales (3)	Video producer (1)
designer/manufacturer (2)	Marketing (2)	Visual artist (15)
Community arts worker (3)	Media presenter, journalist,	
Curator (1)	producer (7)	

Separate to recorded interviews, organisers of festivals in Darwin were surveyed using a postal questionnaire. The bulk of questions in this were from a similar questionnaire used to assess the economic and cultural significance of festivals in southern States on the ARC Rural Festivals Project (DP0560032; see Gibson *et al.* 2009). That one chief investigator (Gibson) was present on both projects led to a similar method being used in the two projects. The Darwin survey had extra questions on how remoteness affected festivals; on Aboriginal input into festivals; and on whether proximity to Asia was beneficial. Of 39 festivals identified in Darwin and sent surveys, 13 were completed and returned (by a combination of music, environment, film, visual arts, food, community, gay and lesbian, and gardening festivals). As with interviews, it is not possible here to include a full breakdown of results from this survey. Interview material constitutes the bulk of our analysis around key themes of remoteness and proximity, with occasional integration of findings from the festival survey where appropriate.

The curse of remoteness

Several interviewees reported on the unsurprising truth that Darwin is distant from key centres of the creative industry, and thus is not well connected to important gatekeepers, opportunities, touring networks and sponsors. According to an arts manager,[2] Darwin's remoteness from the Australian eastern seaboard was 'a weakness, when you're in a national table and the needs of the Territory and parts of WA, Queensland, and South Australia—the more remote areas—are not necessarily given enough reference in developing some of the policy' (interview, July 2007). Simple matters of critical mass and distance combined to make efforts to support arts activity difficult:

> It's not that accessible, we don't have the opportunities for the rate bases in our urban areas to support arts development, we don't have a lot of the benefits that the eastern seaboard, or South Australia, or even Perth have in terms of their demographics, their numbers, their accessibility ... we just don't have the population, we just don't have the same working conditions. We have escalated costs in arts delivery; even a tour costs a fortune. Artback told me a very funny story about having to hang artworks on fence posts, because there was nothing else. It's hard with those key players on those national tables, for them to understand what the actual working conditions are for our artists. They hear the words but they don't understand, and they're sick of hearing the words, because at the bottom line they go: 'you've got two percent of the nation's population, come on, who cares!' (Interview, April 2007)

Distance from key infrastructures and poor critical mass meant higher costs. Servicing Darwin's 'local' market was more costly, because that 'local' market was made up of far-flung remote communities, rather than more densely interconnected, bigger places. For a music promoter:

> The only reason people don't come here is it's too expensive to bring five people on the plane when they're only going to play a few shows. The only people that tour here are people that come in their own bus and play the markets: two shows a day at the markets and stay for three weeks and play

every show there is in the dry season. They're the only people that can really afford to, or people that are brought up for a one off for the Darwin Festival. (Interview, August 2007)

Festival organisers were asked how remoteness affected overall costs. Every one of the 13 festivals surveyed agreed that remoteness did increase their costs, although they were divided on the extent of this. Four said that remoteness affected the overall cost of staging the festival to a small extent; five said that remoteness affected them quite a lot; and three said that remoteness affected staging to a large extent. The most common costs incurred were transportation for imported exhibition materials and performers; and supplies for larger festivals (with local catering and staging businesses unable to meet their demands, meaning festivals had to look interstate).

Remoteness was also problematised by interviewees because it fostered cultural cringe, and made locals defensive. On the one hand, for a public art manager:

> It's the sense of trying to be the latest thing at the Venice Biennale, when you should just be yourself, I think that's a disadvantage. There's the potential to value things from elsewhere, rather than valuing what's right under your nose, and that's a real disadvantage. (Interview, April 2007)

On the other hand, for a visual artist, remoteness manifest in a tendency of benefactors to demand that commissioned art needed to have a local hook—rather than facilitating artistic expression in whatever its format:

> In Darwin they like the work to be about Darwin ... if you were in Sydney you wouldn't be saying, 'oh we want to make art about Sydney'. It's something that's a regional insecurity, [whereas] my art has a more universal thing. There's nothing in Darwin that particularly interests me, there's no real interesting materials to be found and I'm just not really interested in the vista and beautiful sunsets. (Interview, July 2007)

In Darwin, the danger of parochialism is ever-present (see also Andersen, this issue).

The pleasures of isolation

Isolation and small size combined to reduce connections to international networks. For a key figure in organising cultural festivals, 'a lack of international and interstate influences is big' (interview, June 2007), a sentiment repeated throughout many interviews. However, this kind of statement was often made by interviewees in the form of a juxtaposing sentence: 'I thought that remote *qua* negative ... BUT ... remote/proximate *qua* positive.' The perils of remoteness were quickly contrasted against positives, or affirmations about how Darwin was in turn proximate to other places and attractions. In pointing to the positives, interviewees may well have been 'true believers'—committed to Darwin despite its limitations. Or perhaps creative workers were rationalising trade-offs to themselves (an extension of decisions to remain in Darwin rather than moving to southern States where connections and opportunities might be more widespread). Statements about the pleasures and benefits of isolation were broadly of three types: affirmations of the colonial/frontier context as a source of inspiration; claims about the benefits of working independently in an isolated place; and distance from national and international artistic trends as enabling comparatively laissez-faire artistic expression.

According to a film producer:

> A lot of people come up here just because it's the frontier still, the old myth of it being the badlands or the outback and all that sort of stuff, which is great. In the arts we use that, because it's a big part of our identity and that's fine. (Interview, July 2007)

For a curator and visual artist, creativity was facilitated in Darwin because of its isolation and its proximity to nature—and because, in contrast to conventional wisdom in creative industries research, daily face-to-face interactions with other artists were absent:

> Darwin's strength as a creative city is … the smallness of the place and that it seems that nature is right on your doorstep. It doesn't seem like that in other large cities. So, just being able to stretch your eyes across the harbour and go down to the beach and go for walks and see animals and birds around you, it's really important; that sense of isolation. When I was in Melbourne, there was a lot of visual stimulation; a lot of exhibitions around; everything seemed to come at me, just too many openings to go to, so many things to do. I like that in Darwin I can seek out my interests without being bombarded. I like that isolation. I feel like I can ground myself and have a stronger sense of personal identity without being part of something larger, not doing work like other people around you. (Interview, July 2007)

Indeed, remoteness was frequently celebrated by interviewees as creative 'freedom', distant from metropolitan trends, fashions and compulsions:

> In some ways I see our isolation as an advantage … Those artists who really focus on their work with a level of integrity, and don't look at trying to be the latest thing, actually develop a practice that is grounded in something that is extraordinary, which is here. (Interview, June 2007)

These kinds of sentiments were linked to romantic ideals of the artist toiling in creative isolation. For the manager of a local dance company:

> I think a large number of Australian artists who are developed and work in larger cities, tend to look very broadly, relatively esoterically about their art forms. That really could exist anywhere, as long as there's a black box that you can sort of hide in to do it. One of the things that makes Darwin special, culturally, is that isolation, that you're not influenced by what that critic says, or what that company says. (Interview, April 2007)

For a visual artist:

> You can actually work in relative isolation in Darwin, so your work's not too influenced by what's going on. I think that what's happening in Darwin is a lot of artists are creating work that's raw, and you need vitality, it's not being pasteurised, or trying to pander to somebody else's ideas and theories, or aesthetics. (Interview, July 2007)

The local creative industry scene is eclectic. Hybridisations of urban, outback, traditional Aboriginal, tropical and contemporary Asian styles (in art, sculpture, music architecture) are made possible in Darwin (see below). In addition, if less obviously, small population size limits the possibilities of distinct 'scenes' forming around specific creative activities. While Melbourne might host autonomous creative communities around specific styles of music, subcultures, visual arts movements or design trends, in Darwin creative workers swap styles and locations, work in multiple arts categories, and network across diverse groups—as a survival strategy in a small town. For this reason it is not possible to theorise our results by creative sector—because rarely did creative workers stick to one genre or pursuit. For a musician:

> The location and the size of the city doesn't allow for really strong cliques of people to form. You tend to interact with all sorts of people. Say if I was in Melbourne or Sydney and I was into electronic music, I could hang out with my electronic music friends, but there's just not the critical mass to do that here. So, I make weird electronic music, I'm in a rock band, a reggae band, I play in a Gamelan ensemble, there's so many different outlets that Darwin just exposes you to. (Interview, April 2007)

Creative identity-hopping enables fledgling artists to survive, and reflects the fluidity and comparative openness of cultural expressions in remote, small places.

Fictions of distance: postcolonial proximities

Perceptions of Darwin as remote from southern centres of creativity were contrasted against affirmations of Darwin's local cultural mix—its proximity, in other words, to Asia and major regions of Aboriginal culture and language. For Aboriginal creative workers, Darwin simply was not remote—it was home, central to their universe, networks, positionality. For an Aboriginal photographer:

> I was very confident in Darwin. I was always told from the minute I was able to understand that this is your country, it doesn't matter who else is here, you belong here and so it wasn't even a thing to think about; it was just something that I knew as I grew up going through high school. I always feel totally comfortable in Darwin. (Interview, September 2007)

Darwin is a major urban centre at the heart of a string of networks between remote communities, as evident when places of work outside Darwin were mapped (see Figure 1). The resulting 'spider-web' of connections between remote Aboriginal communities and Darwin illustrates a different kind of regional proximity at work.

For some non-Aboriginal creative workers, the possibility of Indigenous encounter was an attraction, sustaining a lineage of exploration of the frontier experience with echoes of novels such as Xavier Herbert's *Capricornia*, Alexis Wright's *Carpentaria*, feature films such as *Walkabout* and rock albums such as Midnight Oil's *Diesel and Dust* (and even further back, to Rousseau and Gauguin). There was for some a 'raw vitality' of creativity in the tropics, at the edge of colonialism. The dance company manager described how:

> when I originally decided to live here, it was certainly having that face to face contact with Indigenous Australia, which I didn't feel I'd ever had

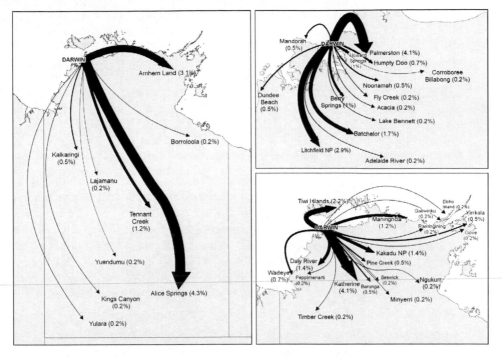

FIGURE 1. Remote community networks: places of work cited in interviews with creative workers, 2007–08, as percentage of total number of places of work cited. Maps courtesy of Chris Brennan-Horley.

before in my life, and I know, for me, that changed my view of what I was as an artist. I thought I knew what it meant to be an Australian, but certainly, coming up here and having that contact with Indigenous Australia, I have totally had to completely wipe out all those things that I thought I was and reassess what it actually did mean to Australia—to be an Australian. (Interview, April 2007)

Indeed, isolation as authenticity has fed the critical reception of Top End Aboriginal art and music in the national and international press—words like 'pure' and 'original' infuse descriptions of Aboriginal artistic practice and music; as evidenced in reviews of Geoffrey Gurrumul Yunupingu's highly successful 2008 album *Gurrumul*—which emphasised his 'pure voice', his uncorrupted music about his life and country (in the Territory's Top End region—not far from Darwin). Aboriginal art and music is made 'authentic' through perceptions of the geographical distances involved in its production and distribution—as with world music more broadly—and a degree of legitimacy stemming from perceptions of disconnectedness from the machinations of urban capitalism (see Dunbar-Hall & Gibson 2004). Awareness that this critical distance might be achieved in Darwin was a genuine reason why some creative workers moved there. More simply, too, proximity to local Aboriginal creative workers fuelled new kinds of expedient, hybrid design, as an architect with an award-winning local firm explained:

We have great connections to the Indigenous culture up here. Part of what we're doing on the entertainment centre is we've commissioned

Maningrida [community Aboriginal art centre] to do a major part of the public artwork ... We're able to access that skill and expertise and the artists quite easily; they're two minutes away at the centre; there's a really easy communication line. So you build on these relationships. That's what we've been trying to do, is build relationships with the artist community, bringing in their expertise into what we do; to add a layer of interest and a layer of detail, and a layer of what we think is more about the Territory, instead of importing materials from down south, or getting anything from China ... the buildings all started looking the same in Sydney, Melbourne. (Interview, April 2007)

Proximity to Aboriginal culture was overlaid with discussion of Darwin's proximity to Asia. For a visual artist, 'instead of looking down there [to southern States], we tend to look to our nearer neighbours for our inspiration and our stimulus' (interview, July 2007). According to a key figure at the Darwin Theatre Company 'the proximity to Asia makes a difference ... Timor's just over the water really; it's closer to us than anything else. It's quicker to get to Timor than it is to Katherine, which is the next town down the road' (interview, April 2007). The General Manager of a major Darwin festival described its role in reinscribing an imagined geography of Darwin:

We're reflecting our geographic position. We sit in Darwin and draw a circle around, we kind of try and programme work from there, so we look at work from Indonesia and Singapore and Papua New Guinea and some Pacific islands as well. (Interview, April 2007)

This amalgam of cultural proximities in a small city produced a setting for creative activity cherished by many workers. For a visual artist:

Darwin's too small for one culture to isolate themselves. In southern cities you're able to just live in one suburb and never leave it, and not have to mix too much with other people, but Darwin's never been like that. Asian people, white people, Mediterranean and European people, Aboriginal people, all mix in and grow up together. It's not something that you really think about until you go somewhere else.

The dance company manager emphasised that:

We have a unique cultural diversity here which has a different presence to other cities. What I'm talking about is our closeness to Asia, our closeness to Indigenous culture ... We're not a ghettoised community. If I look at my friend's network there's a whole heap of people from different cultures that's very natural for most people in Darwin. I think from a creative perspective, access to that is totally inspiring. What does it mean to be an Australian artist who actually lives in Asia, who actually lives really close to the original Indigenous people of this country? I feel very fortunate in that. (Interview, April 2007)

For some creative workers, building networks into Asia was a simple matter of increasing audiences and sales. Darwin was a cost-effective place to trade into Asia. A graphic designer reminded us that:

there are thousands of tourists a year that come from Japan and Asia, specifically to purchase Aboriginal art, and they travel out to communities like Maningrida. When I was at the grand final at Tiwi Islands, they also had the footy art sale, and people were going home with trolley loads full of prints. (Interview, March 2008)

For a visual artist:

Darwin's geographical position is paramount … the geographical placement of Darwin is the thing that most appeals to me … it's so close to Asia. I can travel to Jakarta in a matter of hours for a relatively affordable price which makes my portfolio as an artist in an Australian context, more, almost more domestic in a sense, that it's more controllable by me. It's more common for my colleagues in Sydney or Melbourne to think about going to Asia as a major event, so they would have to get funding; I can fly on Tiger Air for a weekend to simply visit friends.

[Interviewer: And do you gain inspiration from those travels that you would then bring back to Australia?]

Not so much inspiration, I gain market. (Interview, June 2007)

For the dance company:

Apart from anything, it's also more viable financially for us to bring some of their [Philippine] artists to here to work with us, for us to travel; it's hugely cheaper than us bringing someone from Melbourne or Sydney to Darwin. (Interview, April 2007)

Yet for all the talk of proximity to Asia, and of multiculturalism as assets, there was a somewhat uneven translation of this into actual practices. Linkages and networks with Timor, Indonesia and the Philippines took time and patience; some measures of success had been achieved (notably for dance companies and for the Darwin Festival), but for many Asia was still viewed in terms of opportunities and general talk than in real or specific activities. For 6 of the 13 surveyed festivals, proximity to Asia was said to have no positive influence at all. For the public art manager, 'I think one advantage that is very rarely taken up is that proximity with Southeast Asia. It's starting to happen, but I think that not enough happens in that direction' (interview, April 2007). One impediment repeatedly cited was the problem of gaining support funding from federal and Territory arts agencies for international linkages, given that the arts has been considered a marginal 'cultural' rather than economic development concern: 'it's easy to fund someone from Australia to go to do a residency, but if you're one of our artists here who'd like to bring back a large performance, perhaps from Bali, to be part of our festivals and working with the schools, it's difficult to attract funding' (arts worker, interview, July 2007). One theatre company started:

a relationship with a company in Timor and we're going to do a co-production. To do this I have to find corporate funding because I can't use NT arts funding. I've looked into arts funding from other sources like the Australia Council but there's no real mechanisms for us to co-produce with another company in another country. (Interview, April 2007)

Despite many taking advantage of Darwin's tropical proximity to Asia, national borders, assumptions that creative industries are 'cultural' rather than commercial, and the limits of public funding have prevailed.

Conclusions

> The broad cultural mix that you get here, as an artist, I find that incredibly stimulating, and as an Australian, I find that a really interesting place to explore—what does it mean, who am I, you know, who am I in that context? (Local dance company, interview, April 2008)

This article contributes to an emerging geographical literature in creative industries research which extends from 'local' case studies to open out and engage with relational networks of trade and touring, and the circulation of people, goods and ideas across scales (Coe & Johns 2004; Kong *et al.* 2006; Rantisi *et al.* 2006; Waitt & Gibson 2009). Darwin is remote, and yet also occupies a liminal space, on the edge of one large continent, and thus proximate to another; a place grappling with its (recent) colonial legacy and uncertain future. Insights from this case study show how multiple, sometimes overlapping relational geographies of remoteness and proximity jostle for discursive space in the lives and perceptions of creative artists (cf. Ridanpää 2005). Remoteness matters for everyone in Darwin at one level, but is also a state of being in which people unevenly invest or believe; everyday life and work is always perceived in relation to other points of reference (wherever they might be).

Distance can contribute significantly to costs in terms of importing materials and talent, and to exporting them. Touring exhibitions or performers is not a cheap undertaking from Darwin. Singapore or Jakarta is as cost-effective as Melbourne, which can create its own unique opportunities. For some creative practitioners the relative distance from other centres was a boon; it allowed them to develop creatively, free to pursue their own vision, unfettered by the demands of metropolitan trends, fashion and funding imperatives. For others, the isolation was a professional isolation, where one yearned for the stimulation of a competitive environment.

While to much of the rest of the country Darwin appears far away and remote, for the creative practitioners who live there the distance from the Australian eastern seaboard also meant proximity to Asia (and the rest of the world). Indeed, while distance was clearly an issue for many, especially those with family interstate, in an age of relatively affordable air travel, not to mention communication and mail order via the Internet, the idea of remoteness 'as lack' was rare, and Darwin's unique proximities were quickly emphasised. Remoteness is as much a state of mind as a geographical reality.

Acknowledgements

This project was funded as an Australian Research Council Linkage Project (LP0667445; 2005–2009). We extend thanks to the wider research team—Tess Lea, Donal Fitzpatrick, Chris Brennan-Horley, Karen Hughes and Francesca

Baas-Becking—for their various contributions to the project from which this paper stems.

NOTES

[1] Aware of the debates about what constitutes the 'creative' industries, and of the risk of arbitrariness in including or excluding certain sectors (Cunningham 2002; Markusen *et al.* 2008), we adopted a catholic approach to what 'creative industries' might be in Darwin. The British government's definition proved a useful starting point—including advertising, architecture, the art and antiques market, crafts, design, designer fashion, film and video, interactive leisure software, music, the performing arts, publishing, software and computer games, television and radio (www.culture.gov.uk/creative_ industries/default.htm). Further augmentation ensued, with categories and occupations added to mesh with the Australian Bureau of Statistics' Framework for Culture and Leisure Statistics (ABS 2007).

[2] In order to conform to university ethics regulations, interview quotes have been made anonymous in this article.

REFERENCES

AUSTRALIAN BUREAU OF STATISTICS (ABS) (2007) *Arts and culture in Australia: a statistical overview*, Catalogue No. 4172.0., ABS, Canberra.

BRENNAN-HORLEY, C. & GIBSON, C. (2009) 'Where is creativity in the city? Integrating qualitative and GIS methods', *Environment and Planning A* 41, pp. 2595–614.

COE, N. & JOHNS, J. (2004) 'Beyond production clusters: towards a critical political economy of networks in the film and television industries', in Power D. & Scott A. (eds) *The cultural industries and the production of culture*, Routledge, New York, pp. 188–204.

CUNNINGHAM, S. (2002) 'From cultural to creative industries', *Media International Australia incorporating Culture and Policy* 102, pp. 54–65.

DUNBAR-HALL, P. & GIBSON, C. (2004) *Deadly sounds, deadly places: contemporary Aboriginal music in Australia*, UNSW Press, Sydney.

GIBSON, C. (2003) 'Cultures at work: why "culture" matters in research on the "cultural" industries', *Social and Cultural Geography* 4, pp. 201–15.

GIBSON, C. & BRENNAN-HORLEY, C. (2007) *Creative tropical city*, statistical data analysis report for Darwin City Council, NRETA, and Tourism NT, University of Wollongong, Wollongong.

GIBSON, C., WAITT, G., WALMSLEY, J. & CONNELL, J. (2009) 'Cultural festivals and economic development in regional Australia', *Journal of Planning Education and Research* (in press).

GILL, R. & PRATT, A. (2008) 'In the social factory? Immaterial labour, precariousness and cultural work', *Theory, Culture & Society* 25, pp. 1–30.

JULL, P. (1991) *The politics of northern frontiers in Australia, Canada, and other 'First World' countries*, North Australia Research Unit, Australian National University, Darwin.

KONG, L., GIBSON, C., KHOO, L.-M. & SEMPLE, A.-L. (2006) 'Knowledges of the creative economy: towards a relational geography of diffusion and adaptation in Asia', *Asia Pacific Viewpoint* 47, pp. 173–94.

LEA, T., LUCKMAN, S., GIBSON, C., FITZPATRICK, D., BRENNAN-HORLEY, C., WILLOUGHBY-SMITH, J. & HUGHES, K. (2009) *Creative Tropical City: mapping Darwin's creative industries*, Charles Darwin University, Darwin.

LUCKMAN, S., GIBSON, C. & LEA, T. (2009) 'Mosquitoes in the mix: how transferable is creative city thinking?', *Singapore Journal of Tropical Geography* 30, pp. 70–85.

MARKUSEN, A., WASSALL, G., DENATALE, D. & COHEN, R. (2008) 'Defining the creative economy: industry and occupational approaches', *Economic Development Quarterly* 22, pp. 24–45.

POWELL, A. (1988) *Far country: a short history of the Northern Territory*, Melbourne University Press, Melbourne.

RANTISI, N., LESLIE, D. & CHRISTOPHERSON, S. (2006) 'Placing the creative economy: scale, politics, and the material—the rise of the new "creative" imperative', *Environment and Planning A* 38, pp. 1789–97.

REIMER, S. (2009) 'Geographies of production II: fashion, creativity and fragmented labour', *Progress in Human Geography* 33, pp. 65–73.

RIDANPÄÄ, J. (2005) 'Kuvitteellinen pohjoinen: Maantiede, kirjallisuus ja postkoloniaalinen kritiikki' [The imaginary north: geography, literature and postcolonial criticism], *Nordia Geographical Publications* 34, pp. 13–37.

ROSS, N. (2008) 'The new geography of work: power to the precarious?', *Theory, Culture & Society* 25, pp. 31–49.

WAITT, G. & GIBSON, C. (2009) 'Creative small cities: rethinking the creative economy in place', *Urban Studies* 46, pp. 1223–46.

Multiple Work Sites and City-wide Networks: a topological approach to understanding creative work

CHRIS BRENNAN-HORLEY, *University of Wollongong, New South Wales, Australia*

ABSTRACT   *This paper attempts to further spatial understandings of creative work by focusing on the inherent topology linking workplaces together. Topographical approaches to creative employment are advanced by reflecting on how creative activity is linked and enacted across space. Everyday realities of creative work mean that multiple locations are used (for rehearsal, exhibition, for networking or for performance). It is difficult to ascertain relationships between these places using conventional methods such as mapping census data. Instead, I draw on workplace data taken from a creative industry research project conducted in Darwin, a remote city in Australia's Northern Territory, where qualitative interviews and mental maps were combined. The analysis proffers two key advances. First, mental map interviews conducted with creative workers can yield, on average, a fivefold increase over census data in the number of important, everyday work sites reported by creative practitioners. This means more detail and subtlety can be woven into analysis. Second, a hierarchy of important intra and inter-suburban linkages can be mapped, revealing the city's creative topology and furthering breakdown of the 'creative inner/uncreative outer' urban binary. A topological approach reveals that rather than being CBD (central business district)-centric (which static readings of raw workplace counts per neighbourhood show) creativity is highly interconnected across the city. Such findings bolster the case for reimagining suburbs as vital and functional parts of the creative city. Rather than being typified as secondary to internal-CBD milieus, outer suburbs are highly connected, performing specialised roles in Darwin's creative topology.*

Introduction

In the context of creative industry work, the notion of the singular workplace is outdated. Instead, this paper introduces the concept of creative work topology, a synthetic mapping methodology for combining multiple spaces of creative work with the inherent links forged between them. A topological approach to mapping creative work uses the connections between places for inferring how creative work is situated in the city, rather than relying on topographical comparisons between contiguous spatial units.

Reconfiguring creative industry mapping methods to incorporate topology fits with prevalent arguments about networking in creative city literatures. The concept of networking is regularly espoused, with the driving force behind successful creative work said to be reliance on interpersonal relations bound up in degrees of geographic proximity (Pratt 1997; Scott 2000; Wittel 2001). Geographical concepts like clusters and precincts have prospered, predicated on spatially proximate networks and have been actively promoted as a governance tool for creative industries, attempting to provide a degree of spatial fixity to an elusive, yet economically important, set of activities (Pratt 2004; Mommaas 2004; Gibson & Kong 2005).

A disjunct exists between how creative industries are understood—as embedded in networks, both spatial and relational—and the realities of how they are represented cartographically in research. Creative industry activity maps premised on concentrations of creative workers or firms perpetuate ideas of inner-city dominance (Gibson & Brennan-Horley 2006), with networks and connections between actors in inner-city clusters inferred rather than evidentially revealed (Kong 2009). In a policy sense, this reinforces inner-city clusters as key locations in the creativity value chain, masking the role of other locations in the day-to-day operation of creative activities.

Employment statistics sourced from census data are the most often used data sources for such creative industry mapping, but remain an inadequate means for understanding the multiple ways in which creative work is enacted across space (Brennan-Horley & Gibson 2009). The key issue is that employment statistics are premised on the notion of the singular workplace. Such data may prove beneficial for understanding employment patterns for creative occupations and sectors with higher rates of spatial fixity, such as advertising and marketing, but they cannot encapsulate working patterns for creative practitioners whose workdays (or nights) are predicated on fluidity. The day-to-day movements of a musician are a case in point: managing affairs from a home office, rehearsing with various groups in different studios, recording or performing in different spaces again. Each space remains important to that musician's creative work. Yet if musicians appear in census data (and there is much evidence to suggest that census and firm statistics miss musicians altogether—see Gibson 2002), only one site can be accommodated: whichever site they indicate as their major place of employment during census administration.

Creative industry mapping methods need to incorporate techniques for linking multiple sites of individual creative activity if they are to better represent how aggregate creative work patterns unfurl across the city. The logical conclusion is that empirical data are needed from qualitative research with creative workers (Shorthose 2004), capturing 'missing' data and inserting workers' agency back into creative city mapping.

My case study is a city on the margins, Darwin, the capital of Australia's Northern Territory—a city that perhaps could not be further, spatially or metaphorically, from those often-cited 'creative cities' of the northern hemisphere (Luckman *et al.* 2009). By utilising a novel data set of spatially located work attributes from a remote location, this paper seeks to provide a reassessment of urban creativity—especially for places that may not display attributes of the classic de-industrialising northern hemisphere creative city (densely populated inner cities, main street café culture or gentrified ex-industrial districts). In Darwin, different methods were necessary, and from this case are opportunities to rethink more broadly how creativity connects up parts of the city. As my analysis below reveals, creative workers constructed complex networks across their city, working in multiple locations, making use of suburban places almost as much as they utilise spaces of the inner city. By documenting auxiliary workplaces and linkages, we can build up from individual geographies of creative work to a collective typology, positioning parts of the city within a relational network of creative activity.

A remote, suburban city: whither creativity?

This research stems from a wider research project 'Creative Tropical City: Mapping Darwin's Creative Industries', funded through the Australian Research Council's Linkage Project Scheme (LP0667445; 2005–2009; see also Gibson *et al.* 2010). The project was jointly supported by Darwin City Council, the Northern Territory's Tourism Commission and the Arts and Museum Division of the Office of the Chief Minister of the Northern Territory. Each industry partner was interested in exploring Darwin as a creative place, as an exercise in understanding and perhaps influencing liveability in a city suffering from demographic issues associated with population 'churn' (a quarter of the city's population moves every 5 years) and over-reliance on extractive, defence and tourism industries. The project's three stated aims were to determine the nature, extent and change over time of Darwin's creative industries; interrogating the applicability of national and international creative industry policy frameworks to Darwin; and to identify opportunities for local creative industry transformation.

Darwin is a coastal city on the north-western tip of the Northern Territory. Its small population of only 70 000 resides over 1500 km from the nearest substantial town of Alice Springs. The city's urban form is a function of its tumultuous short European history; destroyed four times since 1890, most recently in the 1970s by Cyclone Tracy. The city has since been rebuilt in a decentralised fashion with expansive suburbs featuring wide roads and low building densities; mostly of the single-storey cyclone-proofed variety. Prior to the economic downturn of 2008, Darwin experienced a residential apartment boom, with many large apartment buildings springing up in the central city, catering primarily to the fly-in fly-out labour market of mining, defence and government workers.

The city's economic base includes employment in the military, government departments servicing the territory, in mining and tourism. The city does support an estimated 2000 creative practitioners, comprising 3.2 per cent of the city's workforce (Gibson & Brennan-Horley 2007)—numerically larger than mining and on a par with tourism employment. Darwin plays a key role in the Indigenous Australian art trade and for Indigenous communities in general, operating as a focal point for the many scattered communities from the surrounding Top End region (Lea *et al.* 2009). However, creative industries have not figured in debates about the city's economic future—instead positioned as 'marginal extras' to iconic the frontier industries of mining and tourism. Better understanding Darwin as creative city was therefore a research aim also directed towards imagining the regional economy differently.

There is much to learn from Darwin's peculiar geography. Working and living as a creative producer in Darwin is influenced by the city's unique mix of remoteness from other Australian cities and its proximity to Southeast Asia (see Gibson *et al.* 2010). This is an interesting case study for examining creative work precisely because it is so different from the cities upon which much of the creative city literature is premised. Darwin does not have a dense inner city with disused industrial housing stock waiting for a makeover as a creative precinct. Because of the city's mix of small population size and low density, the applicability of existing creative city mapping techniques were also called into question (Brennan-Horley & Gibson 2009). Mapping creative workers across varying spatial scales is a common methodology for understanding spatial patterns of creative activity (see Gibson & Brennan-Horley 2006; De Propris *et al.* 2009). However, Darwin's small population, and thus comparatively small raw total numbers of creative workers spread out across a predominantly low-density city, meant that statistical analysis of work locations from the census was never going to provide enough detail for understanding how and where creative work is carried out in Darwin.

Ethnographic methods remain the most plausible means for uncovering cultural contexts of creative work (Gibson 2003), and for understanding historical specificities underpinning spatial location in the creative industries (O'Connor 2004; Kong 2009). Hence they played a key role in the methodologies used by the Creative Tropical City project. Although ethnographers have highlighted the underlying influence of space and place on creative work (Drake 2003), until recently researchers have not attempted to reconcile qualitative methods with the quantitative mapping methodologies regularly enacted in economic geography/creative industries mapping projects (Brennan-Horley *et al.* 2010). Darwin offered an opportunity to try new methods, generating a spatial database of work locations (revealed through ethnography) that could be mapped and analysed quantitatively as an alternative to census data. Ethnographic techniques like interviewing and mental mapping were combined with the spatial analysis capabilities of GIS technology. This mixed-method approach provided a synthetic means for revealing the networked topology of creative work in Darwin.

Mental mapping the creative workplace

A key element of the ethnographic research process in the Creative Tropical City project was a mental-mapping exercise. Blank base maps of Darwin consisting of suburb boundaries and a road network were provided to informants during interviews, for them to graphically indicate responses to questions about the 'where' of their creative working lives (see Figure 1). The interview process was not entirely about the mapping: a small number of mapping questions (five in total) were woven in and around other questions about being a creative practitioner in Darwin. Our aim was to elicit spatially referenced responses to questions about creative working lives (for a detailed examination of how the method was developed and deployed, see Brennan-Horley & Gibson 2009; Brennan-Horley *et al.* 2010).

Interviewees were first asked to pinpoint their home, followed by their place of work. It quickly became apparent during trials that 'place of work' as a singular, distinct location was an outdated concept for all of our trial participants. Their everyday working lives revolved around multiple work sites, depending on their creative involvement. When prompted to indicate a place of work, more than one

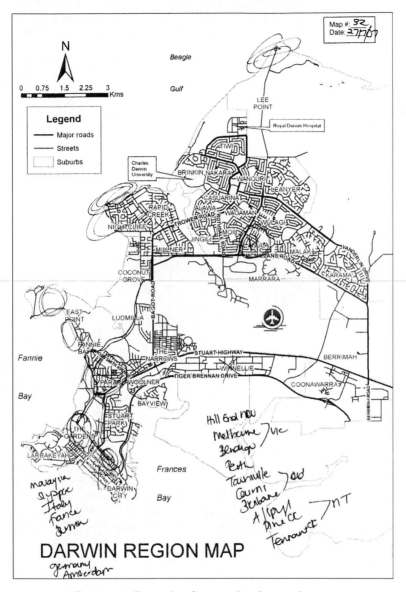

FIGURE 1. Example of a completed mental map.

site was always indicated. In response, our interview schedule was altered, encouraging informants to divulge more detail about what constituted a working day and what sites were involved.

The types and locations indicated were as wide ranging as the workers involved: generated through the mental maps were suburban studios and home offices, rehearsal spaces in industrial zones and inner-city performance sites. Sites for sourcing materials turned up in all manner of places, from National Parks to suburban hardware chain stores (which doubled as a space of networking for some creative industry practitioners). Interviewees were then asked open-ended questions about Darwin's conditions for creative practitioners, including the city's strengths and weaknesses as a creative place and the effects of seasonality and remoteness. Other mental-mapping questions were then woven throughout the remainder of the interview schedule. 'Sense of place'-type questions included asking where participants would go in Darwin for creative inspiration (cf. Brennan-Horley *et al.* 2010), places they like to frequent for recreational purpose and where they felt hotspots of creative activity may manifest in Darwin (Brennan-Horley & Gibson 2009).

The markings drawn on the mental maps were, strictly speaking, geographical approximations. Base maps distributed in interviews were accurate in terms of the locations of Statistical Local Area (SLA) boundaries and the placement of the road network, but they did not contain all identifying information needed to accurately pinpoint locations down to the individual property level (it is questionable whether a map containing large amounts of information is even useful in the interview context). The A3-sized maps used were purposely designed to collect spatial data with a minimal degree of leading—to deter respondents from indicating a facility because it was already indicated for them on the map. There was, however, a tendency for some respondents to place a dot on the textual name of an SLA, rather than seeking the location of the important space within the SLA. This tended to occur more in smaller SLAs, where map scale hindered more accurate spatial responses. We provided informants with paper maps that could not contain all street names. Different map sizes were trialled, each with varying degrees of identifying information, before settling on A3. We were not looking for pinpoint accuracy. Rather, we wanted an approximation, to the best of their abilities, of the parts of the city in which creative work took place and what sort of facility or space was utilised. We could then use this as a starting point for talking about creative lives in relation to these places.

In total, the mental-mapping process yielded 98 maps of creative Darwin from 101 individuals. Ninety-one maps contained data about work sites, in the form of dots or crosses. The missing seven 'non responses' were from informants whose main place of work lay outside the bounds of the Darwin council area, and thus were unable to be captured on the map. These places were written down in the margins and stored separately in the GIS. They were used to produce flow maps documenting connections to regional, national and international locations. While cognisant of the influence these places 'outside' the map play on creative work in Darwin, it was beyond the scope of this analysis to incorporate places beyond the Darwin council area (see instead Gibson *et al.* 2010).

Each work point was geo-referenced using ArcGIS 9.2. The spatial distribution for all work sites is displayed as a density plot in Figure 2. The 200 m radius about each point goes some way to accommodating spatial inaccuracies inherent in hand-drawn locations. Figure 2 reveals creative work aggregations in the SLAs of Inner City, Parap, Nightcliff, Brinkin and Winnellie.

Mental maps yielded 472 work locations, with each informant indicating an average of five locations. When examined at the city-wide scale, this methodology yielded a data set five times greater than that of census counts. This does not imply that Darwin's creative workforce is five times larger but rather that each worker may utilise up to five times as many locations than can be inferred from the census. This alone was a surprising and important finding from this research—enabling much more detail to be mapped.

A key strength of the data source was that each respondent's work locations could be assigned a weighting based on characterisations provided by interviewees.

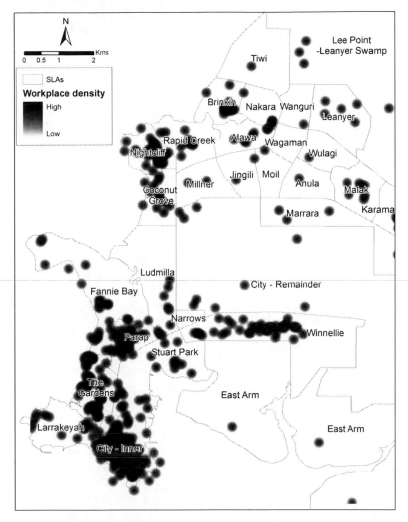

FIGURE 2. Density map of creative workspaces sourced from mental maps (kernel density plot, search radius $= 200$ m, pixel size $= 10$ m^2).

Indicated on each mental map was a primary place of work, represented with a 'W' (see Figure 1). These sites were attributed in the GIS as a 'major workplace'. Ancillary sites marked with a cross were attributed as a 'minor workplace'. Ninety-three work sites were classified by our interviewees as their major workplace, with the remaining 379 points attributed as 'minor'. For every one major creative workplace there were possibly four other sites that had a role to play in that individual's creative practice. Going back to the musician example, here was mappable evidence of their network of work locations: the home office, spaces for rehearsal, recording and performance.

It should be noted that 'ancillary' work sites were not strictly of lesser value in their input into creative work but rather were present along a continuum of increasing importance and frequency of visitation. Returning again to the musician, no greater weighting ought to be given to a site of rehearsal as opposed to that of a performance space—they are equal parts of that practitioner's creative practice. But for an artist, spending time getting work framed at a framing shop or sourcing supplies was considered by most to be of lesser importance than the act of painting. Minor workplaces could perform any number of roles, from the utilitarian, such as sites of supply to more important roles as sites of exchange, networking or performance. The actual usage of each site also depended upon the interviewee: for some a performance venue was a place to perform, for others a site of networking. It is beyond the scope of this paper to drill down to this micro-level of individual locations (though this is the intention of future publications). I turn instead to propose a topological method that can handle the data in an aggregate fashion: a macro-analysis comparing linkages between Darwin's SLAs.

Mapping the networked creative city topologically

With a conceptual basis in set theory and geometry, topology has applicability to a wide range of research areas, including GIS, architecture, computer sciences and cultural research. As conceptualised in computer network visualisation, topology provides a means for graphical representation of relations between nodes in a network. Network topology maps are most frequently used for visualising phenomena in an abstract graphical fashion, but when it is advantageous to visualise the data cartographically, network nodes can be pinpointed in geographic space with connections rendered between them (Eick 1996; Huffaker et al. 1998).

In applying topology to understanding spatial formations of cultural work, I use the term as a descriptor first for understanding that creative work is situated within a network of related sites, and second for visualising the key relational networks between places. A city's creative economy could be envisaged as a construct of relations between sites while retaining an appreciation for the relative location of its constituent nodes. In other words, a creative work typology is cognisant of creative work locations *and* of the number and strength of connections between places where work occurs.

By applying geographical techniques and GIS methods developed for the study of population movements (Tobler 2003; Goodchild & Glennon 2008), the topology of Darwin's creative work sites can be mapped. As each workplace (either major or minor) was indicated spatially by respondents, when processing each mental map a link was established between the primary workplace and its minor partners. Each site was aggregated to the relevant SLA, and SLA-level links were used to populate an in–out interaction matrix. Finally ArcGIS 9.2 was used for geographic visualisation.

Flow mapping permits relationships between places to be mapped in a number of ways. The two most useful for mapping creative linkages were: in gross form, counting connections in both directions between nodes and summing together to give a magnitude; or in a two-way form, consisting of separate lines for the direction of each flow, with the summed magnitudes of each direction equalling the gross flow magnitude.

The city was analysed in the following manner: first, gross flow was determined for the entire city (see Figure 3). To continue to unpack and test notions of inner-city primacy (Gibson & Brennan-Horley 2006), the inner city was carved off and dealt with on its own, mapped first as a gross trend (see Figure 4)—detailing all flows between the CBD (central business district) and suburban SLAs. Second, the 'in-flow' from the suburbs to the CBD was mapped (see Figure 6). In-flow scores represent instances where major workplaces in non-CBD suburbs maintained links with the CBD. Arrows aid in displaying the directionality of the flow relationships, always pointing from major towards minor workplaces. Conversely, the out-flow from the CBD was mapped, displaying the flow from major CBD workplaces to their minor counterparts (see Figure 5).

The remaining SLAs were then analysed. Again, gross flows between suburban SLAs were mapped, showing broad patterns in the suburbs (see Figure 7). The suburbs were then split and mapped into their corresponding two-way flows (see Figures 8 and 9).

Results

A summary of total linkages and breakdowns by in- and out-flows for the CBD and the suburban SLAs is provided in Table 1. It summarises flow totals across four spaces: city-wide, city-inner, intra-CBD and inter-suburban. At the total city-wide level, 377 linkages were indicated on mental maps. When aggregated to the SLA level and related through the in–out matrix, 141 separate SLA-to-SLA flows result.

The linkage with the greatest magnitude is an intra-SLA linkage within the CBD itself, with a magnitude of 55. For the 34 primary work sites within the CBD, a further 55 movements were evident within the CBD in an auxiliary capacity. Given the concentration of facilities for performance, networking, rehearsal and office functions, it is not surprising that the CBD contains the largest internal flow magnitude. Interestingly for Darwin, the CBD was the only SLA that recorded a significant intra-SLA flow, further evidence for the primacy of CBD spaces with their dense interrelated sites and networked creative workers, leveraging off spatially proximate facilities and relationships.

Figure 3 displays the spatial arrangement of gross linkages across the city. The directionality of each linkage cannot be implied, but it does provide an overall visualisation of the dense web of creative connections crisscrossing the city. Results in Table 1 and in Figure 3 imply that strong links occur between the CBD and the remaining suburban SLAs. To better visualise this relationship, the gross relationships between the CBD and the suburbs are displayed in Figure 4.

Figure 4 indicates that there is not a strong concentric pattern forming about the CBD (within which one would expect most CBD-to-suburb links to be with nearby inner-city SLAs, gradating downwards in number, and outwards geographically, to

TABLE 1. Total flow counts and magnitudes, delineated and by city/suburb boundary

Linkages	Flow count	Flow count (%)	Flow magnitude	Flow magnitude (%)
City-inner in-flow	15	10.6	72	19.1
City-inner out-flow	20	14.2	84	22.3
City-inner intra-flow	1	0.7	55	14.6
Gross city-inner subtotal	36	25.5	211	56
Suburban SLAs in-flow	90	63.8	137	36.3
Suburban SLAs out-flow	15	10.6	29	7.7
Gross suburban SLAs subtotal	105	74.5	166	44
Gross Darwin grand total	141	100	377	100

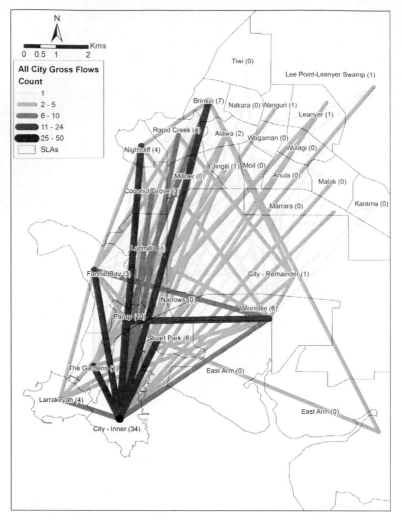

FIGURE 3. City gross creative flows. Numbers in parentheses equal the count of major workplaces within the SLA.

a minimal number of links with outer suburbs). Instead, the results reveal that some of the strongest linkages are forged between the CBD and further-flung sites at Darwin's northern end. Brinkin and Nightcliff both score as highly as other more CBD-proximate SLAs like Fannie Bay, the Gardens and Parap. For the creative practitioners in our study, spatial proximity to Darwin's inner city was not a precursor for forming or maintaining creative work links with sites in the CBD. To better understand the relationship between the CBD and the suburbs more fully, the directionality of the flows between the CBD and suburban SLAs was then further broken down.

Figure 5 takes the gross city-wide link data and displays only the number and magnitude of flows occurring between major CBD work sites and their ancillary suburban counterparts. Twenty links between the CBD and particular suburban SLAs result, accounting for 25.5 per cent of all SLA linkages across the city. Their collective magnitude equals 22.3 per cent of Darwin's total creative topology.

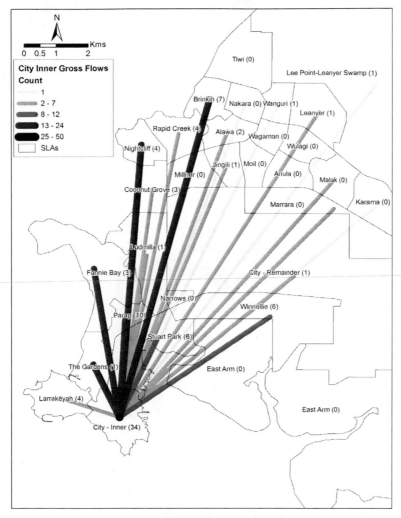

FIGURE 4. Inner-city-to-suburban SLA gross flows. Numbers in parentheses equal the count of major workplaces within the SLA.

The strongest link was maintained with the Gardens (17 links). As the name suggests, the Gardens is dominated by Darwin's botanic gardens and houses the world famous Mindl Beach Markets (whose offices are CBD based). Market culture is a key feature not only of Darwin's creative economy but of Darwin generally. Various markets around the city function as key consumption sites for artisan crafts, as performance spaces for live music and also as spaces for networking. In addition, the Gardens are home to the annual Darwin festival, a key date on the city's cultural calendar and a festival of national and international stature. This strong link between the CBD and the Gardens is testament to the influence of climate on Darwin's creative economy (Luckman *et al.* 2009; Brennan-Horley *et al.* 2010). The dry monsoonal winters guarantee the viability of outdoor spaces for creative consumption, hence the vibrant market and festival culture.

Other strong links between the CBD are also prevalent: with Parap (10 links), a burgeoning local gallery district and, again, site of a weekly (albeit more

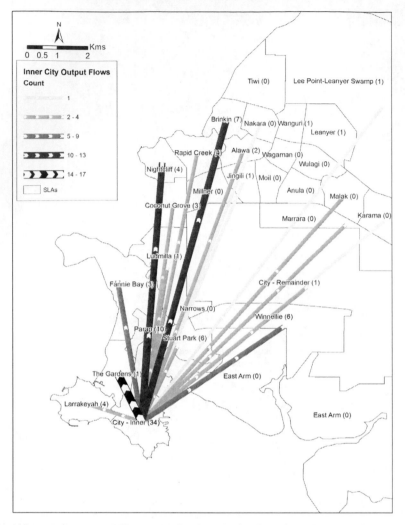

FIGURE 5. Inner-city output flows to suburban SLAs. Numbers in parentheses equal the count of major workplaces within the SLA.

community-focused) market (Brennan-Horley & Gibson 2009); with Nightcliff (10 links), which also has a local market, but also houses key rehearsal spaces like Corrugated Iron Dance Studio; with Brinkin (10 links), home of Charles Darwin University; and with Winnellie (7 links). These are important suburban nodes supporting primary workplaces of the inner city. Figure 6 depicts the reverse relationship: major creative workplaces in suburban locations maintaining ancillary links with CBD sites. In total there are 15 SLA-to-CBD connections with a combined magnitude of 72, accounting for 19.1 per cent of all flows across Darwin's creative topology.

The strongest SLA-to-CBD link occurs between businesses based in Parap, with 14 links to the CBD stemming from the 10 creative practitioners headquartered here. Stuart Park, Fannie Bay and Brinkin all had between seven and nine linkages. Not surprisingly, there was little evidence of a reverse relationship between the Gardens and the CBD to that observed above. Primary work sites in the Gardens

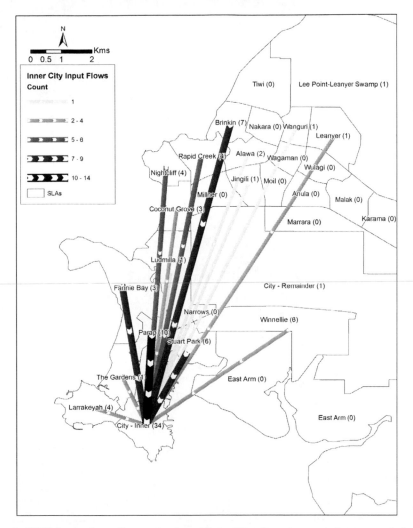

FIGURE 6. CBD input flows from suburban SLAs. Numbers in parentheses equal the count of major workplaces within the SLA.

only maintain two links with minor workplaces in the CBD. Given there was only one creative business situated here as a major place of work, it is understandable that the flow direction is stronger from the CBD to Gardens rather than the reverse.

The linkages in Figures 5 and 6 illustrate that ancillary suburban workplaces are proving, in magnitude, to be of a greater importance to the functioning of creative workplaces in the CBD than ancillary spaces in the CBD itself. They constitute 22.3 per cent of overall CBD linkages, compared to only 14.6 per cent of intra-CBD connections.

Separating the CBD invites further analysis into the way the remaining suburban linkages are configured. Figure 7 maps gross flows between suburban SLAs in a non-directional format, highlighting the overall strength of creative linkages between the suburban SLAs. Shown in a gross form, certain suburban SLAs are functioning as hubs of interaction for creative work in Darwin's suburbs. Winnellie maintains strong links with Parap, with Fannie Bay and Nightcliff. It also maintains

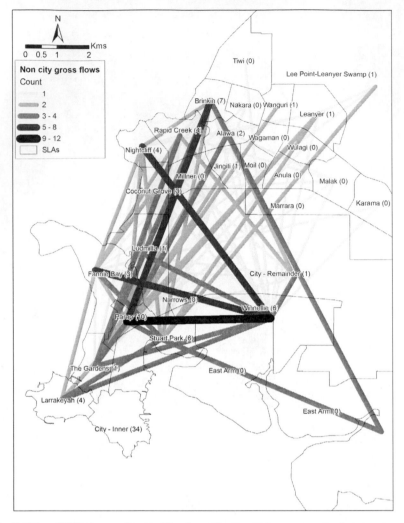

FIGURE 7. Non-CBD gross flows. Numbers in parentheses equal the count of major workplaces within the SLA.

a web of smaller links with other SLAs. Similarly, Parap maintains links with Winnellie but also Brinkin and a web of other SLAs.

In total, 105 linkages were evident between suburban SLAs, accounting for 75 per cent of all links across Darwin. Their combined magnitude equals 44 per cent of the city-wide total: nearly half of the creative linkages across the Darwin council area occur between suburban locations. The strongest SLA-to-SLA relationships appeared between Winnellie and Fannie Bay/Parap/Nightcliff and between Brinkin/Parap.

As with the earlier maps, the gross flows can be broken down into their comparable in/out component flows to obtain a sense of directionality. The suburb flow data were not mapped in a one-to-many style, as with the maps of the CBD (one) to suburbs (many). Doing so would become unwieldy, resulting in 30 separate maps, one for every suburb. Instead, suburb flow data were mapped in a many-to-many style, teasing apart the gross lines in Figure 7 and revealing any dominant flow directions in the data. Figure 8 maps input-flows across Darwin's non-CBD SLAs. Directional arrows along each line help to determine from which SLA the flow emanates. In total, the suburban SLA input flows accounted for 36.3 per cent of all city-wide linkages.

The strongest links were evident between major workplaces in Winnellie and Fannie Bay, from Charles Darwin University in Brinkin to the gallery spaces of Parap. Apart from the strong linkages in Figure 8, there was also a fine web of connections representing singular links between suburbs. When added to the stronger links represented in darker shades, Parap (44), Winnellie (41) and the Gardens (22) all appear to function as key creative work nodes for suburban creative practitioners. Although not as plentiful as CBD out-flows in Figure 5, there were 12 out-flow linkages occurring between a handful of suburban SLAs. These linkages are mapped in Figure 9, with the strongest occurring between major workplaces in Parap, linking to ancillary locations in Winnellie.

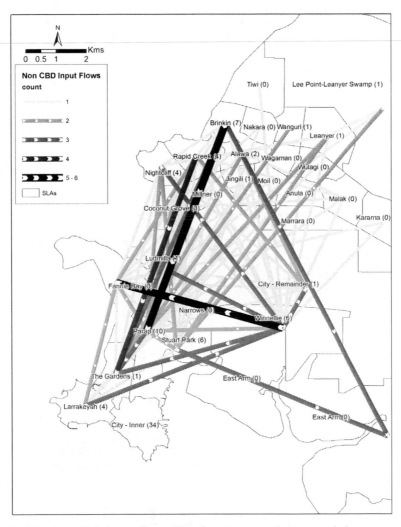

FIGURE 8. Suburban SLA input flows. Numbers in parentheses equal the count of major workplaces within the SLA.

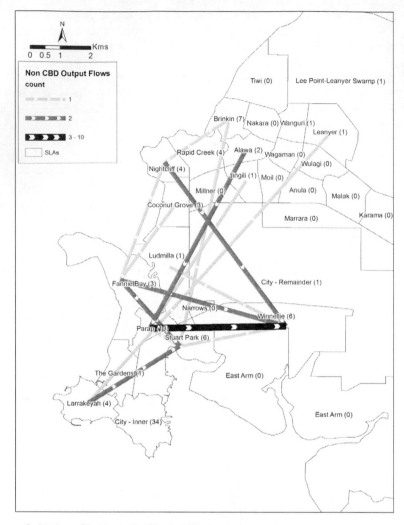

FIGURE 9. Suburban SLA output flows. Numbers in parentheses equal the count of major workplaces within the SLA.

Unlike two-way linkages between the CBD and the suburban SLAs, suburb-to-suburb linkages tended to be more mono-directional. The majority of suburban SLAs served specialised functions within Darwin's creative economy, while the CBD offered a wider range of facilities and evidently stronger connections to workers located in the suburbs. But, based on lines of connection in Figure 9, suburban SLAs such as Winnellie and Stuart Park do provide ancillary services to major workplaces located in a small number of suburban SLAs.

Overall, the CBD remains an important space for creative work, by both number of major workplaces found there (34) and as a site for ancillary usage by major workplaces located in the suburbs. Eighty-five per cent of our sample utilise the city in their daily working lives, meaning only 15 per cent of creative practitioners interviewed did not use the CBD in some sense. But the flow maps show that the role the suburbs play in the creative economy does need reimagining. Linkages between suburban SLAs (45 per cent) accounted for close to half of all connections across the city, and linkages between CBD-based major work sites and ancillary places in the suburbs were greater in magnitude than intra-CBD flows. The picture is of suburban importance and specialisation within a more complexly networked topology integrating as well as bypassing the central city.

Conclusion

Network topology maps provide a strong evidential base for repositioning the suburbs as much more than simply a dormitory location for inner-city workers. Instead, the magnitude of networked relationships that involve the suburbs implies that the role of the suburbs is vital to the functioning of the creative economy. Such findings are of special significance in regional or remote locations, places of low population or low density, where opportunities for networking or available facilities and infrastructure are scarce and distributed over larger areas. Creative practitioners across the city are still networking and sharing resources, but in places like Darwin they are forced to make do with less, to fan out and source materials and utilise spaces that are on offer regardless of their location, rather than working and remaining in distinct spatial clusters or milieus.

Reflecting on the method, mental mapping provides a means for documenting creative work where conventional mapping methods may be less forthcoming. As was the case in Darwin, straightforward readings of census data reproduced a familiar story of inner-city dominance at the expense of 'hidden' creative suburban networks (Lea *et al.* 2009). Mapping the variety of locations that enable creative practice through ethnographic interviewing yielded a more thorough appraisal of the geographies of creative work. Furthermore, understanding topological linkages between spaces provides a framework for understanding how, at the neighbourhood scale, parts of the city function together as an interlinked and networked entity.

Such methods may potentially prove beneficial for all creative cities, especially as inner-city areas become overtly gentrified, homogenised and over-priced for those involved in informal creative sectors and pursuits—fuelling outer-suburban movement in the creative industries (Indergaard 2009). The methodology used here can assist creative city researchers and policy makers to understand the role suburbs play in the creative city for all manner of creative practitioners and, importantly, can convey to what degree the suburbs relate to centres and to each other. Creative work topologies provide evidence for creativity away from overtly-badged 'creative precincts', as well as indicating to what degree inner-city spaces rely on their suburban counterparts.

REFERENCES

BRENNAN-HORLEY, C. & GIBSON, C. (2009) 'Where is creativity in the city? Integrating qualitative and GIS methods', *Environment and Planning A* 41(11), pp. 2595–614.

BRENNAN-HORLEY, C., LUCKMAN, S., GIBSON, C. & WILLOUGHBY-SMITH, J. (2010) 'GIS, ethnography and cultural research: putting maps back into ethnographic mapping', *The Information Society* 26(2) (in press).

DE PROPRIS, L., CHAPAIN, C., COOKE, P., MACNEILL, S. & MATEOS-GARCIA, J. (2009) *The geography of creativity*, NESTA interim report, available from: http://www.nesta.org.uk/assets/Uploads/pdf/Research-Report/Geography-of-creativity.pdf (accessed 5 September 2009).

DRAKE, G. (2003) ' "This place gives me space": place and creativity in the creative industries', *Geoforum* 34, pp. 511–24.

EICK, S.G. (1996) '3D geographic network display', *Computer Graphics and Applications 16*, pp. 69–72, available from: http://www.visualcomplexity.com/vc/project_details.cfm?id=77&index=77&domain= (accessed 20 January 2009).

GIBSON, C. (2002) 'Rural transformation and cultural industries: popular music on the New South Wales Far North Coast', *Australian Geographical Studies* 40, pp. 336–56.

GIBSON, C. (2003) 'Cultures at work: why "culture" matters in research on the "cultural" industries', *Social and Cultural Geography* 4, pp. 201–15.

GIBSON, C. & BRENNAN-HORLEY, C. (2006) 'Goodbye pram city: beyond inner/outer zone binaries in creative city research', *Urban Policy and Research* 24, pp. 455–71.

GIBSON, C. & BRENNAN-HORLEY, C. (2007) *Creative Tropical City: statistical data analysis report for Darwin City Council, NRETA, and Tourism NT*, University of Wollongong, Wollongong.

GIBSON, C. & KONG, L. (2005) 'Cultural economy: a critical review', *Progress in Human Geography* 29, pp. 541–61.

GIBSON, C., LUCKMAN, S. & WILLOUGHBY-SMITH, J. (2010) 'Creativity without borders? Re-thinking remoteness and proximity' *Australian Geographer* 41(1), pp. 25–38.

GOODCHILD, M.F. & GLENNON, A. (2008) 'Representation and computation of geographic dynamics', in Hornsby, K. & May, Y. (eds) *Understanding dynamics of geographic domains*, CRC Press, Taylor & Francis, Florida, pp. 13–27.

HUFFAKER, B., JUNG, J., NEMETH, E., WESSELS, D. & CLAFFY, K. (1998) 'Visualization of the growth and topology of the NLANR caching hierarchy', *Computer Networks and ISDN Systems* 30, pp. 2131–39.

INDERGAARD, M. (2009) 'What to make of New York's new economy? The politics of the creative field', *Urban Studies* 46, pp. 1063–94.

KONG, L. (2009) 'Beyond networks and relations: towards rethinking creative cluster theory', in Kong, L. & O'Connor, J. (eds) *Creative economies, creative cities: Asian–European perspectives*, Springer, Dordrecht, pp. 61–75.

LEA, T., LUCKMAN, S., GIBSON, C., FITZPATRICK, D., BRENNAN-HORLEY, C., WILLOUGHBY-SMITH, J. & HUGHES, K. (2009) *Creative Tropical City: mapping Darwin's creative industries*, Charles Darwin University, Darwin.

LUCKMAN, S., GIBSON, C. & LEA, T. (2009) 'Mosquitoes in the mix: how transferable is creative city thinking?', *Singapore Journal of Tropical Geography* 30, pp. 70–85.

MOMMAAS, H. (2004) 'Cultural clusters and the post-industrial city: towards the remapping of urban cultural policy', *Urban Studies* 41, pp. 507–32.

O'CONNOR, J. (2004) ' "A special kind of city knowledge": innovative clusters, tacit knowledge and the "Creative City" ', *Media International Australia* 112, pp. 131–49.

PRATT, A. (1997) 'The cultural industries production system: a case study of employment change in Britain 1984–91', *Environment and Planning A* 29, pp. 1953–74.

PRATT, A. (2004) 'Creative clusters: towards the governance of the creative industries production system?', *Media International Australia* 112, pp. 50–66.

SCOTT, A.J. (2000) *The cultural economy of cities*, Sage, London.

SHORTHOSE, J. (2004) 'Accounting for independent creativity in the new cultural economy', *Media International Australia* 112, pp. 150–61.

TOBLER, W. (2003) 'Movement mapping', unpublished paper, available from: http://csiss.ncgia.ucsb.edu/clearinghouse/FlowMapper/MovementMapping.pdf (accessed 8 February 2009).

WITTEL, A. (2001) 'Towards a network sociality', *Theory, Culture and Society* 18, pp. 51–76.

Making Connections: creative industries networks in outer-suburban locations

EMMA FELTON, CHRISTY COLLIS & PHIL GRAHAM,
Queensland University of Technology, Kelvin Grove, Australia

ABSTRACT *The role of networks and their contribution to sustaining and developing creative industries is well documented. This article argues that although networks operate across geographical boundaries, particularly through the use of communication technologies, the majority of studies have focused on the ways in which networks operate in (a) specific inner-urban metropolitan regions or (b) specific industries. Such studies are informed by the geographical mindset of creative city proponents in which inner-urban precincts are seen as the prime location for creative industries activity, business development and opportunity. But what of those creative industries situated beyond the inner city? Evidence in Australia suggests there is increasing creative industries activity beyond the inner city, in outer-suburban and ex-urban areas. This article identifies characteristics of creative industries networks in outer-suburban locations in Melbourne and Brisbane drawing on extensive ethnographic fieldwork. It argues that supporting and sustaining creative industries networks in these locations may require different strategies than those applied to inner-city networks.*

Introduction

The creative industries are commonly associated with innovation and entrepreneurism. For organisations that are SMEs (small and medium enterprises), spatially based local clustering is considered important for networks to facilitate knowledge integration and business development (Florida 2002; Pratt 2007; Scott 2006). Close proximity of businesses enables networks to operate effectively in a value chain relationship which consists of team members, clients, suppliers and stakeholders (Pratt *et al.* 2007; Adkins *et al.* 2007; Lazaretti *et al.* 2008). Moreover, as commercial sectors that are often dependent upon project-based, contractual work and subject to the volatility of market economies, the creative industries are high-risk industries (McRobbie 2002). Self-employment and job insecurity are common, and the experience of risk is intrinsic to those working in the sector (Banks *et al.* 2000). The management of risk, while acknowledged as a feature of late modernity in general (Beck 1992), impacts on organisations to differing degrees, but is central to the experience of many creative industries workers (Banks *et al.* 2000). According to Kong, such risk is 'often countered by relationships of trust, a form of social solidarity' managed by networks of social relations (2005, p. 64).[1] In this context, networks support not only knowledge, creative energy and industry development, but also function in several other ways for the creative industries worker: as a method for finding continual employment, managing unease related to job insecurity and for enabling new cooperative endeavours (Coe 2000).

This article explores professional networking practices, types, and scales in an often-overlooked field: creative industries situated in outer-suburban Australia.[2] Despite the fact that Australia is highly suburbanised, there is little research on the creative industries workforce in suburban locations. Given that the topography and culture of outer-urban areas is distinct from inner-urban regions, with different infrastructure, amenities and population distribution, what are the implications for networking activity? As networking is critical for business development, how does it work in low-density outer-suburban areas? Can effective creative industries networking occur in geographical sites which are neither dense 'precincts', 'hubs', nor 'clusters'? While technology enables a virtual space for communication and collaboration, creative industries literature repeatedly points to the importance of locality for the types of creative industries and creative industries networks which emerge in specific sites (Gertler 1995; Bathelt *et al.* 2004; Lange *et al.* 2008). Most creative industries network studies tend to focus on inner-city sites, where the focus of creative industries policy and analysis remains (Kong 2005; Pratt *et al.* 2007; Wittel 2001). This focus is explicable: much creative industries activity is indeed urban, and the key drivers of economic growth as defined by classical economic geography—'production resources, skills, and institutions of coordination'—often concentrate in urban areas (Storper & Scott 2009, p. 64). To date, there has been little research that addresses the impact of outer-suburban local places on creative industries, or on the impact of creative industries on outer-suburban places (for exceptions see Gibson & Brennan-Horley 2006; Brennan-Horley, this issue).

As urbanisation intensifies, redistribution of land use and zoning continues to change the morphology and use of cities and their suburbs. The ways in which people live and work across many suburbs has changed significantly in the period of urban consolidation since the 1980s. For instance, only 14 per cent of jobs are located in Melbourne inner-city and central business districts, with the remaining jobs located in suburban and outer-suburban localities (Davies 2009). Because creative industries are 'embedded in networks and institutions that are socially-constructed and culturally-defined' (Coe 2000, p. 394), an understanding of the specific geographies and networks in which they are embedded provides insight into the factors that shape the development of the sector when located some distance from the urban core (see also Brennan-Horley, this issue).

The shifting suburbs

Australian outer suburbs are no longer (if they ever were) strictly locations of domesticity and a retreat from working life. The outer suburbs now tend to be places of demographic plurality and social and economic complexity (Gibson & Brennan-Horley 2006; Gleeson 2002; Randolph 2004; Salt 2006; Turner 2008). Processes of urban consolidation have contributed to structural shifts in the suburban landscape—low-density suburbs have been in-filled and mixed-use development has enabled the growth of large-scale multipurpose buildings, changing the structure and experience of the suburbs (Gibson & Brennan-Horley 2006, p. 456). Consequently, what people do and how they live in the outer suburbs have changed. Studies of creative industries in the outer suburbs in Australia's largest city, Sydney, find that ex-urban Statistical Local Areas (SLA) such as Wollongong and the Blue Mountains experienced the highest rates of creative industries employment growth in Sydney between 1991 and 2001 (Gibson & Brennan-Horley 2006, p. 467). Similarly, the ex-urban areas of Wyong, Camden and Wollondilly achieved higher rates of growth in creative work in the last 20 years than did inner-city SLAs such as Sydney City and Marrickville (2006, p. 465). In part, this is because outer-suburban areas are the fastest growing areas in Australia, so it is not surprising that their creative industries workforces are also growing quickly (Gibson & Brennan-Horley 2006, p. 468). When paired with real estate trends, for many in the precarious employment typical of the creative industries the inner city is simply now too expensive a place in which to work and live.

Method

This research draws on the findings of an Australian Research Council project investigating the experience of creative industries workers in four outer suburbs of Brisbane and Melbourne. This study focuses on two locations, Redcliffe in Brisbane and Frankston in Melbourne, because they present the most activity and share similar demographic and physical characteristics. Frankston has the larger number of creative industries workers at 646, while Redcliffe has 251, according to Australian Bureau of Statistics data (2006).[3] While identifying that creative industries activity relies on quantitative data, identifying and under-standing the operation of networks requires ethnographic and qualitative research methods (Brennan-Horley & Gibson 2009; James 2006). Findings are based on 82 in-depth interviews with creative industries workers. Interviews averaged 45 minutes and used open-ended questioning to canvass a range of issues in relation to networking. While the research is location specific, we argue that the study's findings are indicative of network characteristics that can be applied across creative industries located beyond inner-city regions. This is because the network-ing issues identified are intrinsically geographical; that is, they operate within an outer-peripheral/inner-metropolitan relationship, and in a relationship to the outer-suburban location and the surrounding regions.

This study attends to three types of networks: (1) those developed and maintained amongst participants[4] with colleagues who work in the same industry or creative practice but work in other companies or elsewhere; (2) professional and informal networking associations which contribute to professional development and business opportunities; and (3) those developed between clients and potential clients.

Network analyses identify numerous types of professional networks, and studies such as Coe's (2000) acknowledge that any organisation or individual is simultaneously involved in multiple network types and geographical scales. As Coe argues, professional networks span across international, national, regional, and local boundaries (2000; see also Grabher 2002). A comprehensive network analysis would identify all types of networks—and might employ Brown and Keast's (2003) 'continuum of connectedness' to define the strength of each network type—and would also identify the organisation's embeddedness in various geographical scales of network. But this article focuses on a diverse industrial sector—the creative industries—rather than on a single organisation or a single industry, and thus a comprehensive network analysis is not feasible. While we acknowledge the importance of regional and international networks, a focus on the local adds insight into the growing literature which attends to 'creative places' (Kong *et al.* 2006; Luckman 2009; Luckman *et al.* 2009). Locality-based ethnographic observation provides insight into less formal aspects of networking, through observations about where people met, who met in those venues and what types of encounters and exchanges took place. Ethnographic observation and informal encounters in various venues within the locality enabled some degree of understanding about how tacit knowledge was shared by participants in each location.

Definitions of creative industries vary across regional jurisdictions and national boundaries. This article uses the Creative Industries National Mapping Project's (CINMP)[5] six-category definition of creative industries, which is itself derived from the British Department for Media, Culture and Sport's foundational definition:[6] film, television and entertainment software; writing, publishing and print media; advertising, graphic design and marketing; architecture, visual arts and design; music composition. Yet in conducting the research, two identifiable and distinct groups emerged from the larger entity of 'the creative industries' as defined by the CINMP: these groups are shorthanded in this article as 'commercial' and 'artisan' creative workers. While it is clear that there are overlaps between the two groups, it is useful to differentiate between the two in order to account for the complex dynamics of the creative industries: as Drake notes of creative industries research, 'it is important that empirical research focuses on specific sub-sectors ... [in order to] reflect the enormous diversity and eclectic character of the creative industries as a whole' (2003, p. 516). Some studies perform this detailed accountancy by focusing solely on one sector—Pratt (2002), for example, focuses on the new media sector, and Kong (2006) and O'Regan and Ward (2008) attend to the film industry—while other studies attend to the entire 'creative industries' sector (e.g. Luckman 2009), accounting for its broader dynamics. Like Drake (2003), this article uses two groupings of creative industries workers in order to highlight differences and similarities between the two. 'Commercial creative workers', for the purposes of this article, are defined as multimedia designers, graphic designers, architects, advertising workers, entertainment software designers, and publishers. 'Artisans' here comprise the categories of writers, musicians, visual artists, illustrators and performing artists. While these are not standardised categories in the field of creative industries research, they are simply a refinement of the existing CINMP definition—a refinement driven by the empirical data themselves—and they thus articulate to the wider scholarly field (see James 2006 for discussion of the importance of transferability and dependability in cultural economic geography).

Parallel to the business development model of networking is the relationship between networks and sociality. Kong's (2005, p. 73) study of networks in the Hong Kong film industry identifies the ways in which networks impinge on the local communities in which they operate, emphasising the connection between sociality, culture and economics. Creative industries networks, in other words, both rely on and add to a community's social capital. The symbolic value of culture and creative industries more broadly, in contributing to the symbolic value of place identification, is widely acknowledged (Zukin, 1995; Harvey 1992). Redcliffe and Frankston are both bayside suburbs with a strongly articulated sense of local identity, and both suburbs have well-established arts-based communities. The creative workers in this study living and working in these suburbs tended to be known throughout their local community, either through their businesses or as local artists. In tacit ways, their presence added to the local identity and to the symbolic capital of place (Brecknock 2009). The contribution of the creative industries to the symbolic value of their outer-suburban places parallels the similar role that creative industries play in adding to place identity in inner cities (see, for example, Adkins et al. 2007).

Locality and networks

This study's interview data indicate three principal concerns that delineate networking patterns and issues confronted by workers in the outer suburbs. One of the key findings, corroborated by other studies with creative industries participants trans-geographically (Wittel 2001; Kong 2005), is the value placed on face-to-face interaction, despite participants' wide use of technology (see also Gibson et al., this issue). This had broader ramifications in several ways, but most importantly for the solvency and development of businesses for commercial creative workers. We have grouped the three concerns as: (1) distance from the inner city, (2) technology vs personal contact, and (3) places for networking.

(1) Distance from urban centre

> I used to belong to the Design Institute of Australia but then I just found I wasn't getting enough out of it . . . I think a lot of the events are very city-focused. I don't think it caters for businesses outside of that central city hub.
> (Tracy[7], graphic designer, Frankston)

Distance from the inner-city's CBD (central business district) was articulated most commonly by commercial creative participants as an obstacle to attending networking events organised by professional associations. Most participants also acknowledged that distance from the CBD limited the potential growth of their business, through lack of informal networking opportunities. The location of their businesses in outer-suburban localities entailed trade-offs.

Redcliffe and Frankston are both situated approximately 40 km from their capital city centres of Brisbane and Melbourne, linked by arterial roads which make accessibility by car to the city or regions beyond relatively easy. Public transport to Frankston is good, with train and rail connections. Redcliffe has greater limitations with only a bus and no train connection. Among the design-based industries of architecture and advertising, 37 per cent of participants cited the relative ease of driving to the city via the highway and cheap airfares as a positive feature enabling them to meet with clients and to attend occasional city-based networking opportunities. However, the majority of design-based SMEs—75 per cent— regarded distance as both a perceptual and geographical barrier for connecting with professional networking organisations; and for the artisan group, for showcasing their work at city venues which attract wider audiences. The barrier created a two-way negative relationship between the inner city and the outer suburb. For outer-suburban workers the distance made attending formal professional networking activities less likely and appealing; it also meant that clients and perhaps more importantly *potential* clients from the inner city were less likely to conduct business with one located in the outer suburb. Furthermore, the perception of the outer-suburban locality as an obstacle in attracting city-based clients by potential clients was articulated by several participants, as graphic designer Jacinta observes:

> we have had calls up in the city region but then quite often I find because people discover we're in Frankston, because we don't advertise that on our website, when they discover we're down here sometimes they shy away.

Devoid of the symbolic cachet and amenity of inner cities, both Redcliffe and Frankston were seen to suffer from image problems. Although both localities have established arts communities and are attractively situated on bays, they do not rate highly on the cultural radars of their respective cities. Both localities endure somewhat negative legacies informed by historical associations with the suburbs' traditionally low socio-economic profiles and with the historical Australian antipathy to suburban communities (Kinnane 1998).

Not surprisingly, SME participants recognised there were more significant business opportunities in the inner city, and that networking there would facilitate greater work opportunities. Yet the cost of locating a business in the inner city was far greater than the costs of running a business in the outer suburb. This recognition was met with compromise, informed by both *lifestyle* and *lifecycle* concerns. In relation to lifestyle, participants had weighed up the financial, physical and other costs of business location, and reached a compromise in which a trade-off between income and lifestyle was acknowledged. In their outer-suburban locations, participants felt they could enjoy a more relaxed and, for some, family-friendly lifestyle (see also Luckman 2009), while accepting the financial limitations for business development. Participants who observed and commented on this trade-off tended to be middle-aged, and many of them prioritised family considerations (see Verdich, this issue). Indeed, as Mommas observes, 'it is not the case that all members of Florida's ominous "creative class" prefer the conviviality of the inner-city. Some groups ... prefer to live and work in more homogenous suburbs' (2009, p. 53). The extent to which establishing a creative industry in an outer suburb is both a lifestyle and lifecycle motivation is illustrated by advertising director John, now based in Redcliffe. In his early 50s, John ran a successful inner-city-based agency for two decades, while living in and commuting from Redcliffe. Keen to purchase business premises rather than renting, his outer suburb offered more affordable property than the inner city. By purchasing his premises John was able to obtain greater financial security, and, increasingly important at this stage of his life, time for leisure and family:

> ... when you get to my age you think you may as well make the last five years comfortable years rather than the stress. Every time I go to meetings in the city, because we still go to our client meetings you know ... you sit in traffic and you think yeah, I know why I moved.

The value of the bayside locations of Redcliffe and Frankston was frequently cited as enabling creativity, particularly among the artisan group, but with recognition of the financial trade-off, as Grant, a Frankston illustrator, notes:

> I'm more creative here but I would get more work because of the contacts if I lived in the city or near the city ... I miss a lot of networking ... the distance impacts on your work ... if I lived closer to the city, you might find people dropping in or they know you're close so you can pop around ... and you talk about work and you may get a job.

Grant reiterates what is already widely known about the clustering effects of creative industries workers, that spatial proximity is critical for new knowledge creation and business growth. In many outer-suburban areas where creative industries are nascent, with a widely and thinly distributed population, coupled with business premises that tend to be scattered throughout commercial precincts and industrial estates, obstacles to networking are greater than for creative workers in the high-density inner city.

Several participants articulated the obstacles to attending professional networking events in terms of time and distance, as these are inevitably based in the inner city. Advertising manager John stated that when he moved his company out to Redcliffe: 'it was difficult to go to monthly meetings and things like that. So I do most of my networking through client groups.' Similarly, Gavin, a Frankston-based illustrator, identifies the potential job losses associated with not attending networking events:

> I miss a lot of networking. I missed one last night and I missed one last Friday ... Friday night traffic and I'd been working in a school all day, I was too tired to make the trip. But had I made the trip I'd have met the publisher of ABC books, the new publisher, and could have got a job out of it, but I couldn't make it. So the distance ... impacts on your work.

The importance of face-to-face networking for the acquisition of potential clients is a recurrent theme among creative commercial participants. Without the critical mass of people with whom to network, new business opportunities are more limited than if their SMEs were based in the inner city. Most participants maintained some networks in their outer-suburban locality, although these were seen to have limited value for creative workers. Although all participants had attended events organised either by local councils or business organisations, part of the perceived limitation was the lack of understanding or interest in the different ways in which commercial creative people worked. Judy, a fashion designer in Redcliffe, articulated the difference between her approach to business and those of a local business networking group:

> ... they don't work the way I work basically ... everybody had to stand up and say what their five year plan was and what they do ... and how they got to this point. Very business focussed. I'm not in the slightest like that so when I stood up and said well actually it grows depending on my clientele and depending on where I want to take it. So when I stood up and said that's how I work, I was shot down.

The lack of focus on creative industries business styles and models made local business networking activities potentially dull for several participants, coupled with the fact that participants did not see the value in developing further networks within the localities because of their small scale. Simon, owner of a Redcliffe-based graphic design SME, said that while he enjoyed being connected to the local community through informal networks, he'd been disappointed by his lack of success at obtaining local start-up business grants through his council.

The artisan group, by comparison, tended to be very well networked within its communities and surrounding regions, maintaining several types of networks. This group is connected through local galleries, markets and arts-based practice groups and is perceived to add symbolic and cultural value to its communities (Brecknock 2009).

(2) Personal contact vs technological communication

Technology's capacity to bridge spatial boundaries is frequently at the forefront in discussions of contemporary business development and creative industry networking activity. Questions are raised about the relevance of place in a borderless, global network of technological relations which enables communication and the conduct of business online (Castells 1989). Creative industries such as multimedia, graphic design and advertising are heavily reliant on technology for production, distribution and communication. Despite an increasingly globalised knowledge economy in which technology plays a central role, claims that technology has foreshadowed the death of geography have been laid to rest. Research in the fields of new suburbanism, creative industries and networking activities has reasserted the power of place (Gertler 1995; Kong 2005; Scott 2001; Thrift 1994). One of the points here is that the spatial clustering of producers is seen to facilitate 'unstable, finely grained, frequent and mediated face-to-face contact' in which self presentation, face-to-face negotiating and interpersonal skills are crucial (Thrift in James *et al*. 2006, p. 10). Moreover, the importance of socio-cultural determinants of economic success and the specific epistemological communities in which each has its own vocabularies, knowledges and practices are factors most effectively transmitted through personal encounter (ibid.).

Research with new media workers and filmmakers in Britain and Hong Kong (Pratt 2002, 2007; Wittel 2001; Kong 2005) indicates that face-to-face contact provides tacit forms of knowledge irreplaceable by technology, and the sociality of encounter is highly valued by creative workers. Thus for creative workers beyond the urban core where spatial clustering occurs less and businesses are more thinly scattered across geographical areas, face-to-face interpersonal networking presents specific challenges. Our research asked participants to what extent technology bridges a spatial divide for creative workers who are located beyond the inner city. To a large extent, the findings corroborate other research on the role of technology and networking (Wittel 2001; Kong 2005): while technology facilitates communication and business practices in many productive ways, it does not enable the same opportunities that face-to-face interaction provides (cf. Warren & Evitt, this issue). For people living and working beyond the inner city, this has specific implications. All participants used technology for communicating with colleagues and doing business with clients, yet the value placed on interpersonal networks was repeatedly articulated. The following comments from Peter, a theatre manager in Frankston, are representative of the views and ways in which participants used technology, with the exception of musicians, many of whom found technology valuable for networking.[8] Peter emphasised that technology did not replace the value of face-to-face interaction: 'you need to get out there and see what they're doing and know what they're about … I mean it's more an informal talking'. Peter uses the example of sharing knowledge by talking with other directors about the details of production design such as 'how did you make the witch melt when you did the Wizard of Oz?', the type of knowledge gained through informal conversations between people with whom there is regular interpersonal contact. Similarly, Greg, an illustrator from Frankston, stated that 'the most information I get on what I'm doing and who to talk to is other illustrators'.

For design-based SMEs, technology was used to maintain contact with clients and to conduct some aspects of business online. The typical pattern among participants in advertising and graphic design was that initial business was conducted with face-to-face meetings where ideas and concepts were pitched, and once a relationship was established, a significant component of work was then conducted online. The significance of personal contact was evident in the ways in which clients were found. Participants across both commercial and artisan groups consistently asserted that they gained their work through personal contact and word of mouth. Tim, an architect in Frankston, echoed what other Frankston- and Redcliffe-based architects said: 'it's all word of mouth. We've always worked on that sort of a basis.' Similarly, for Justin, an architect in Redcliffe 'I don't actively go out and seek work. Basically it comes to me through informal networking and word of mouth.' The extent to which participants networked within their local communities was reflected in the types of clients and in the way business was developed. Redcliffe graphic design SME owner Joshua, who was an active networker, saw an advantage in being situated in a small community with a strong sense of local identity. He observed that the value that people placed on their community worked well for his business because he gained a lot of local support, commenting that Redcliffe was an asset for his type of business. However, views on local business support were not seen this way uniformly across all industries. Redcliffe architect Joanna thought that when coveted large-scale design jobs came up along prime real estate waterfront precincts, developers invariably looked for architects from beyond the outer suburbs. The implication here is that developers have a perception that outer-suburban architects are unable to deliver the quality design of architects located in inner-city practices.

The development of networking activities through personal interaction was more pronounced among artisan groups at both study localities, with evidence of networking activity impacting on the communities in which creative workers live. Most artisan participants were well networked through formal and informal networks, with high levels of personal interaction through their participation in place-based practice and professional support groups—teaching and community activities such as holding positions on local committees and running skills development workshops. Among the artisan participants in Redcliffe was a group of very active networkers who were attempting to positively reshape the identity of the suburb through their creative practices. Predominantly in the 21–35 age group, they have staged several local festivals, and one participant, Samantha, has developed a locally based creative social enterprise which connects older women in the community with younger women to make design-based craft products.[9] Their networking activities have resulted in a type of creative collective; one member, Katrina, describes the motivation and activities of the group:

> My friends and I have an organization called the Red Revolution ... we're trying to cause a cultural revolution in Redcliffe ... to try and get the culture to fester a bit more ... we organize festivals in Redcliffe and we're organizing one fairly soon, all locally produced music and product ...

Tangible outcomes of artisans' networks are evident in place-based activities and events such as those described by Katrina, contributing to the symbolic and cultural significance of their locales. In Redcliffe and Frankston, regular cultural events such as exhibitions at local galleries, markets and festivals are held, and products and performances are displayed and distributed. Cultural markets are connected to place, symbolically and materially, through the production, distribution and consumption of place-based products (Zukin 1995).

(3) Places for networking

Venues or 'hubs', places where interpersonal networking can occur, are paramount for the social dimension of networking in which information, tacit knowledge and relationship building are developed through personal encounters (Gertler 1995). Formal networking events such as those organised by professional associations are largely city based because this is where many of their members are located. Local business networking groups in Redcliffe and Frankston were regarded as having mixed value for participants, but generally their use was limited. In inner-city regions such exchanges often occur in local service-oriented amenities such as cafés, restaurants, galleries and at professional networking events.

In their study of social capital in Adelaide's suburbs, Baum and Palmer (2002) identified a lack of amenities and facilities where people could meet, and a concomitant dissatisfaction amongst residents. Arguably, the limited availability of public and quasi-public amenities such as cafés is a feature of many Australian outer suburbs, unless a large shopping centre exists. This is borne out in Redcliffe, while Frankston has slightly more public facilities such as a TAFE college, which function as creative industries hubs. The quasi-public hubs such as cafés along the waterfront at Redcliffe are not well patronised for networking, perhaps because they are seen as places for tourists or visitors from out of the suburb. Rather, across both research locations, participants used public facilities and commercial premises for informal networking, such as libraries, art galleries, the local TAFE college, community arts hall, and commercial premises such as bookshops, music stores, pubs and sound and dance studios. During fieldwork in Redcliffe and Frankston, we observed many casual networking encounters that occurred in commercial places (see also Brennan-Horley, this issue). Fashion designer Lydia who has a shop in Redcliffe notes how her shop functions as a hub for many locals, wondering whether this would happen to the same extent if her shop were located in the inner city:

> people use this very much as a drop in centre. They always know when I'm getting new things or I'm getting something new. They always know when they come in that it's going to be different. If I was in New Farm (inner city), would that happen? You know, people have probably got a little bit more time on their hands here ...

A lack of diversity, as well as the scale of hubs in comparison with the inner city, means that, for some people, outer-suburban meeting places do not cater for their needs. This may be related to Bourdieu's idea of 'habitus' (1984) in which a set of socialised dispositions enable people to feel more comfortable in certain environments and not others. Factors such as ambience, décor and style of venue convey symbolic meaning and may function as barriers or enablers to participation. For members of the networking group Red Revolution, the lack of appropriate places to meet means that they meet regularly at each other's homes. Samantha, the founding networker, described how their meetings began:

> And we just met them at the local ... hey, come along on Sunday night (to Sam's house) ... so slowly ... we've just been gathering people ... we had these little signup sheets to join our revolution ... and I sent the email out to people so it's just been this slow kind of gathering of people. Yeah, more and more I'm amazed at who does live in suburbia.

The group's weekly networking meetings have produced larger community-based networking events such as markets and festivals, bringing people together from across the Redcliffe peninsula and beyond.

Overall, despite the reduced availability of facilities for networking in the outer suburbs, participants were nonetheless active and resourceful about how and where they networked. The extent to which some groups and individuals pursued networking activities such as the artisan group above, points to the importance of sociality (Banks *et al.* 2000; Kong 2005) and of social trust for creative entrepreneurs.

Conclusion

Networking is a complex activity using multimodalities in which both tacit and codified knowledge is shared and relationships built. While researchers have articulated the importance of networking and spatial proximity for business development, little attention has been paid to networking activities among creative workers beyond inner-city areas. This article lays the foundations for understanding how creative industries networks function at a local level in outer suburbs, demonstrating that the dense proximity of cluster networks of the inner city are not the only environment in which creative industries operate. In the outer suburbs studied in this research, artisan local networks were found to be strong and horizontal (comprised of close relationships between individuals), while local networks at the more commercial end of the creative industries spectrum tended to be weaker, and more vertical (with firms contracting other local firms and drawing on the local labour pool to assist in business processes). While some networks connected the outer suburbs to the inner city, network relationships were not purely the 'hub-and-spoke' configuration assumed in thinking that privileges the inner city as *the* locus of creative industries activity (see also Brennan-Horley, this issue). As this study of the local networks of outer-suburban creative industries demonstrates, the geography of creative industries is more complex than simple concentric-circle models—in which inner cities are the hub of creative industries activity, and in which that activity diminishes with distance from the inner core—suppose. The cultural economic geography of creative industries—and of cities—is more complex, and less spatially concentrated, than much creative industries thinking assumes.

Acknowledgements

The authors wish to thank the Australian Research Council and QUT for their support. Grant code: DP0877133.

NOTES

[1] See Coe (2000) on the difference between social, cultural, professional, and familial networks.
[2] This article focuses on creative industries workers who work in outer suburbia, not on creative industries workers who live in outer suburbia but work elsewhere.
[3] The data are combined with the CINMP categories of creative industries; see p. 63 of this issue for categories.
[4] For ethical reasons, participants' names have been changed.
[5] Higgs *et al.* (2007).
[6] http://www.culture.gov.uk/what_we_do/creative_industries/default.aspx
[7] Names of interview participants have been changed in accordance with ethics guidelines.
[8] Most musicians made use of the Internet networking site MySpace, for collaboration, networking and marketing.
[9] The enterprise Biddybags employs young female designers who create designs for retro-style bags, dresses and tea cosies, and elderly women in the community with craft-based skills such as knitting and crocheting make the products. The profits are split between the two groups of women.

REFERENCES

ADKINS, B., FOTH, M., SUMMERVILLE, J. & HIGGS, P. (2007) 'Ecologies of innovation: symbolic aspects of cross-organizational linkages in the design sector in an Australian inner-city area', *American Behavioural Scientist* 50(7), pp. 922–34.
BANKS, M., LOVATT, A., O'CONNOR, J. & RAFFO, C. (2000) 'Risk and trust in the cultural industries', *Geoforum* 31, pp. 453–64.
BATHELT, H., MALMBERG, A. & MASKELL, P. (2004) 'Clusters and knowledge: local buzz, global pipelines and the process of knowledge creation', *Progress in Human Geography* 28(1), pp. 31–56.
BAUM, F. & PALMER, C. (2002) '"Opportunity structures": urban landscape, social capital and health promotion in Australia', *Health Promotion International* 15(4), pp. 351–61.
BECK, U. (1992) *Risk society: towards a new modernity*, Sage, London.
BOURDIEU, P. (1984) *Distinction: a social critique of the judgement of taste*, Routledge, London.
BRECKNOCK, R. (2009) *Richer than the sum of its parts: a cultural mapping study for Moreton Bay Regional Council*, report for Moreton Bay Regional Council, Queensland.
BRENNAN-HORLEY, C. & GIBSON, C. (2009) 'Where is creativity in the city? Integrating qualitative and GIS methods', *Environment and Planning A* 41(11), pp. 2595–614.
BROWN, K. & KEAST, R. (2003) 'Citizen–government engagement: community connection through network arrangements', *Asian Journal of Public Administration* 25, pp. 107–32.
CASTELLS, M. (1989) *The informational city: information technology, economic restructuring and the urban–regional process*, Blackwell, Cambridge, MA.
COE, N. (2000) 'The view from out West: embeddedness, interpersonal relations and the development of an indigenous film industry in Vancouver', *Geoforum* 31, pp. 391–407.

DAVIES, A. (2009) 'The structure of suburban employment in Melbourne', unpublished doctoral thesis, Faculty of Architecture, Building and Planning, University of Melbourne.

DRAKE, G. (2003) '"This place gives me space": place and creativity in the creative industries', *Geoforum* 34, pp. 511–24.

FLORIDA, R. (2002) *The rise of the creative class*, Basic Books, New York.

GERTLER, M. (1995) 'Being there: proximity, organization and culture in the development and adoption of advanced manufacturing technologies', *Economic Geography* 71, pp. 1–26.

GIBSON, C. & BRENNAN-HORLEY, C. (2006) 'Goodbye pram city: beyond inner/outer zone binaries in creative city research', *Urban Policy and Research* 24, pp. 455–71.

GLEESON, B. (2002) 'Australia's suburbs: aspiration and exclusion', *Urban Policy and Research* 20, pp. 229–32.

GRABHER, G. (2002) 'The project ecology of advertising: tasks, talent, and teams', *Regional Studies* 36, pp. 245–62.

HARVEY, D. (1992) *The condition of postmodernity*, Blackwell, London.

HIGGS, P., CUNNINGHAM, S. & PAGAN, N. (2007) *Australia's creative economy: basic evidence on size, growth, income and employment*, technical report, available from: http://eprints. qut.edu.au/8241/ (accessed 20 March 2009).

JAMES, A. (2006) 'Critical moments in the production of "rigorous" and "relevant" cultural economic geographies', *Progress in Human Geography* 30(3), pp. 289–308.

JAMES, A., MARTIN, R. & SUNLEY, P. (2006) 'The rise of cultural economic geography', in Martin, R. & Sunley, P. (eds) *Critical concepts in economic geography: Volume IV, Cultural Economy*, Routledge, London, pp. 1–18.

KINNANE, G. (1998) 'Shopping at last! History, fiction and the anti-suburban tradition', *Australian Literary Studies* 18(4), pp. 41–55.

KONG, L. (2005) 'The sociality of cultural industries', *International Journal of Cultural Policy* 11(1), pp. 61–76.

KONG, L., GIBSON, C., KHOO, L.-M. & SEMPLE, A.L. (2006) 'Knowledges of the creative economy: towards a relational geography of diffusion and adaptation in Asia', *Asia Pacific Viewpoint* 47(2), pp. 173–94.

LANGE, B., KALANDIDES, A., STOBER, B. & MIEG, H. (2008) 'Berlin's creative industries: governing creativity?', *Industry and Innovation* 15(5), pp. 531–8.

LAZARETTI, L., BOIX, R. & CAPONE, F. (2008) 'Do creative industries cluster? Mapping creative local production systems in Italy and Spain', *Industry and Innovation* 15(5), pp. 549–67.

LUCKMAN, S. (2009) 'Creativity, the environment and the future of creative lifestyles: lessons from a creative tropical city', *International Journal of the Humanities* 7, pp. 1–10.

LUCKMAN, S., GIBSON, C. & LEA, T. (2009) 'Mosquitoes in the mix: how transferable is creative city thinking?', *Singapore Journal of Tropical Geography* 30, pp. 70–85.

McROBBIE, A. (2002) 'From clubs to companies: notes on the decline of political culture in speeded up creative worlds', *Cultural Studies* 16(4), pp. 517–31.

MOMMAS, H. (2009) 'Spaces of culture and economy: mapping the cultural-creative cluster landscape', in Kong, L. & O'Connor, J. (eds) *Creative economies, creative cities: Asian European perspective*, Springer, Amsterdam, pp. 45–59.

O'REGAN, T. & WARD, S. (2008) 'Managing uncertainty: defining the location interests of a greenfield location (with a special focus on the new film and television production studios on the Gold Coast)', in Wasko, J. & Erickson, M. (eds) *Cross-border cultural production: economic runaway or globalization?*, Cambria Press, Amherst, NY, pp. 155–86.

PRATT, A. (2002) 'Hot jobs in cool places. The material cultures of new media product spaces: the case of south of the market, San Francisco', *Information, Communication & Society* 5(1), pp. 27–50.

PRATT, A. (2007) 'The new economy or the emperor's new clothes', in Daniels, P., Leyshon, A., Bradshaw, M. & Beaverstock, J. (eds) *Geographies of the new economy*, Routledge, London, pp. 71–86.

PRATT, A., GILL, R. & SPELTHAM, V. (2007) 'Work and the city in the e-society: a critical investigation of the sociospatially situated character of economic production in the digital content industries in the UK', *Information, Communication & Society* 10(6), pp. 922–42.

RANDOLPH, B. (2004) 'The changing Australian city: new patterns, new policies and new research needs', *Urban Policy and Research* 22(4), pp. 481–93.

SALT, B. (2006) *The big picture: life, work and relationships in the 21st century*, Hardie Grant, Melbourne.

SCOTT, A.J. (2001) 'Capitalism, cities and the production of symbolic forms', *Transactions of the Institute of British Geographers* 26, pp. 11–23.

SCOTT, A. (2006) 'Entrepreneurship, innovation and industrial development: geography and the creative field revisited', *Small Business Economics* 26, pp. 1–24.

STORPER, M. & SCOTT, A. (2009) 'Rethinking human capital, creativity and urban growth', *Journal of Economic Geography* 9, pp. 147–67.

TURNER, G. (2008) 'The cosmopolitan city and its Other: the ethnicising of the Australian suburb', *Inter-Asia Cultural Studies* 9(4), pp. 568–82.

UZZI, B. (1997) 'Towards a network perspective on organizational decline', *International Journal of Sociology and Social Policy* 17, pp. 111–55.

WITTEL, A. (2001) 'Toward a network sociality', *Theory, Culture and Society* 18(6), pp. 51–76.

ZUKIN, S. (1995) *The culture of cities*, Basil Blackwell, Cambridge, MA.

Magic Light, Silver City: the business of culture in Broken Hill

LISA ANDERSEN, *University of Technology, Sydney, Australia*

ABSTRACT   *This article examines cultural industries in Broken Hill—the iconic 'Silver City' of Australian mining in far western NSW—and comes from research funded by arts and regional development agencies during 2006 and 2007. In interviewing and surveying the 'movers and makers' of the local cultural sector a picture emerged of a successful group of mainly informally qualified professional visual artists and crafts people working from home studios who spend more time on their practice and make more money than their metropolitan counterparts. Broken Hill also has a thriving service sector, fine weather and competitive location infrastructure for screen industries, and a community proud of its 'arid artists' and its historical and international reputation as a film set. Artists enjoy the lower-than-city costs of accommodation, the quality of light, their proximity to 'Outback' and industrial landscapes, and sustainable local and seasonal tourist markets. With a focus on richly coloured landscape painting and traditional crafts and some contempt for the city 'art mafia', there is limited diversity of cultural products and a 'half-Sydney' market ceiling price on local sales. The Indigenous arts sector has a low profile and is surprisingly—given high numbers of international tourists—underdeveloped. The arts community is fragmented by divisions that both reflect the male-dominated, rugged independence and 'us and them' heritage of this desert mining and 'union town' and inhibit cooperative development. Remoteness means wariness of newcomers and new ideas; young people leave; limited access to business expertise, production services and training; and high transport costs. Isolation means a unique local culture; a friendly community; freedom from city-based art fads, stress and busyness; and blue skies, time and a clear view.*

Imagine a place completely burnt up, not a green leaf or blade of grass, not a particle of shelter, and the heat of the sun about 170 degrees; add to this dust such as I never experienced (enveloping the place like fog), and you may perhaps realise my position in having to sketch in the open. Must also tell you that the ground was so terribly hot that all the time I had to keep picking my feet up, for it felt like standing on hot coals.

71

Brothers of this Brush in England, how would you like to sketch under these conditions? (From an 1888 letter by artist Edmund Harral, Broken Hill Historical Society)

I strongly believe this is the artistic capital of—not just Australia—the Southern Hemisphere. (Interview with a Broken Hill artist, 2007)

Introduction

This article discusses cultural industries in and around Broken Hill in far western NSW and stems from research undertaken in 2006 and 2007 through the Far Western Regional Development Board, NSW Department of State and Regional Development, Regional Arts NSW and West Darling Arts. Here I draw on findings of that research project to address the specific themes of this special issue—the impacts of remoteness and isolation on creative production.

Located on the semi-arid Barrier Ranges—traditional country of the Wiljakali people—the City of Broken Hill has a population of 20 210 (ABS 2008), observes Central Standard Time, and is 1166 km from Sydney. The history of 'Silver City', as Broken Hill is sometimes called, has been dominated by the mining of one of the world's largest deposits of silver, lead and zinc. The surrounding region consists of large properties and small, isolated communities connected by sealed and unsealed roads. The city is a regional centre for the 16 million ha West Darling pastoral industry and the surrounding towns of Menindee, Tibooburra, White Cliffs and Wilcannia.

Mining began in 1883 with the first lease on the Line of Lode—marked by a 'broken hill' noted by Charles Sturt in 1844—and the Broken Hill Proprietary Company (BHP) was incorporated in 1885. The 1933 census recorded Broken Hill with 26 975 residents, the third largest urban population in NSW behind Sydney and Newcastle. The insularity, 'rugged independence', 'making do' and 'make our own fun' characteristics of extreme remoteness—water was carted to the city until the opening of a pipeline from Menindee after the Second World War—combined with the technical skills and tools of the mining industry and what Julaine Allan described in 2008 as 'mining's relocation culture' have had a dominating influence on social memory and cultural practice. Alongside this is the powerful heritage of the local labour movement, as noted in the Broken Hill City Council's 2005 Cultural Plan:

> Mining and the related trade unionism have been central to Broken Hill's culture and have influenced all aspects of life in the City. Apart from the specific impacts of unionism ... relationships between members of the community, trade unionism and socialism spawned many of Broken Hill's cultural icons.

It has been a town where

> civic and community affairs have been dominated by workers and their unions ... where workers ... achieved a degree of collective control unparalleled in Australian historical experience. (Ellem & Shields 2000, p. 116)

It led to the establishment of NSW's first regional art gallery in 1904, followed in 1907 by the first regional public library.

There is a strong local feeling of 'us and them' in Broken Hill. It has an accompanying terminology—if locally born you are an 'A grouper', if you marry a local you qualify as 'B grouper', and all others are from 'Away'.

From the 1940s onwards the population of miners slowly began to fall as a result of declining mineral stocks, changes in demand and increasing industry automation, and by the 1970s the Line of Lode was no longer the most important mining field in Australia. For the past 30 years global zinc prices have driven boom and bust cycles in the local economy—currently in downturn during the global financial crisis with mine closure and the loss of 440 local jobs in mid-2008.

Heritage and outback tourism has become increasingly important for the local economy and—with 18.5 per cent of the population over 65 (ABS 2008)—social services for a 'retirement town'. The regular turnover of health, mining and education professionals from Away, who leave with a 'souvenir' of their time in Broken Hill, and the growing number of tourists—currently around 9000 international and 140 000 domestic overnight visitors per year (Tourism Research Australia 2008, p. 1)—created a market for local arts and crafts. In the past 40 years the arts and cultural sector has played an important role in income generation, location appeal, entertainment, heritage and the creation of local cultural identity.

Locally born arts identities who left to pursue their practice elsewhere include: actor Chips Rafferty, who returned to make his last film, *Wake in Fright*, in the region in 1971; singer June Bronhill, who took her stage name from the town; comedian and author Steve Abbott, aka The Sandman; and musician 'Lord' Tim Grose, who formed the heavy metal band Dungeon in Broken Hill before heading to Sydney. But visual artists, including Pro Hart, Jack Absalom, Badger Bates and Albert Woodroffe, have been more inclined to stick around and this is reflected in the demographics of local arts practice. Pro Hart was one of Australia's most popular and commercially successful visual artists and sculptors and he lived in the Far West all his life. This former miner and famously self-taught and self-promoting artist—who felt that the metropolitan 'art mafia' never accepted him as a 'proper' painter—developed a hobby into international reputation from the 1960s, painting the region with a richly coloured, naïve style. Germaine Greer, writing in *The Guardian* in 2006, described his landscapes as

> not just illustrations of outback life: they glow with the unforgettable light
> of the inland. His gangling twisted feather-top trees are portraits of the
> acacias and casuarinas that refract the raking sun of the desert edge in a
> luminous haze ... The syncopation in the replication of their gnarly boles
> is the genuine rhythm of the Murray-Darling.

During the 1970s, Pro Hart and fellow Broken Hill artists Jack Absalom, Eric Minchin, John Pickup and Hugh Schulz—a group that emerged from the hobbyist Willyama Art Society—formed the 'Brushmen of the Bush' and collaborated on more than 50 Australian and international exhibitions, establishing a reputation for Broken Hill as a centre for 'Outback art' and arid landscape painting. Interviewed in the *Sydney Morning Herald* of 26 May 2006, John Pickup said:

> We wanted to show the rest of the world what life in our home town was
> like. The quality of the light. The magnificence of the country. And the
> unbelievable colours of the soil.

On his death in 2006, Pro Hart was awarded the first state funeral for a visual artist and the first held in western New South Wales.

The passing of Broken Hill's best-known resident gave rise to a sense in some that the local arts community was declining and ageing. West Darling Arts and the Far Western Regional Development Board initiated a research project to gain a better understanding of the 'state of the arts' in Broken Hill and to assist in improving the business climate for local cultural industries. The research describes a confident, growing sector of older artists who earn more income and have more time to work on creative practices than their metropolitan counterparts. The report made 51 recommendations for developing the sector and increasing community engagement, including a professional artist network, a market development agency, mentoring and apprenticeship programs for young people, greater visibility for Indigenous arts, and the establishment of a film museum in the town (see Andersen & Andrew 2007).

Method

While understanding that creative making—particularly in rural and remote regions—will range along a non-profit to for-profit continuum of activities, the research was concerned specifically with 'practising', 'professional' 'creative' practitioners living in the region. As Throsby and Hollister have noted:

> The *practising* aspect means that we confine our attention to artists currently working or seeking to work in their chosen occupation. The term *professional* is intended to indicate a degree of training, experience or talent and a manner of working that qualify artists to have their work judged against the highest professional standards of the relevant occupation. (2003, p. 14; my emphasis)

For the Australian Bureau of Statistics:

> *Professional creative* participants are broadly defined as creative partici- pants who have a serious commitment to their arts practice and consider it a major aspect of their working life, regardless of their income or employment status. (2006, p. 6; my emphasis)

During the cultural audit phase, categories of creative practice and support roles of specific interest in Broken Hill were identified as:

1. Visual arts, craft and design.
2. Music and performing arts.
3. Traditional and Indigenous practice.
4. Writing and communication.
5. Community cultural development.
6. Support roles—teachers, cultural managers, screen industry services, produc- tion suppliers and arts retailers.

The project was launched by the local Member of Parliament at Broken Hill Regional Art Gallery and achieved extensive local media coverage. To build trust, the Far Western Regional Development Board also widely distributed a newsletter that introduced the researchers. A Research Advisory Group—made up of artists, arts business owners, arts support workers, local, federal and State government officers, and community development workers—was consulted throughout the project. To ensure that some long-standing divisions within the local arts sector did not negatively impact on the research, members were asked to encourage their friends and colleagues to participate.

Existing information proved surprisingly difficult to identify and source, and was often not locatable. The main issues were: information on projects or programs was not recorded and there was an absence of tracking and evaluation processes; significant losses of community knowledge and momentum when people 'moved on'; and the politics of previous projects, where lack of action taken, factionalism or disagreements, or a sense of failure led to an unwillingness to record information and a desire to 'forget'. This may be a generic issue for researchers working on cultural industries in small and remote places (see Lea *et al.* 2009). One community development worker observed that:

> A lot of ideas in Broken Hill seem to stall—they get to a certain point and then don't go any further. Sometimes it is because the people who have been part of development and championing leave and the idea just stops. Sometimes the message you get ... as soon as you say that 'I have got this idea' is 'they tried it elsewhere and it failed' ... Because other projects have failed I don't think it's a reason to stop all projects—you need to learn from the mistakes and not make them again.

A cultural audit was completed—including searches of telephone books, tourism literature and the Web, recommendations from cultural managers and artists, and observations from 'walking the streets'—and a database (now housed with West Darling Arts) of local professional and emerging artists, cultural businesses and relevant support services was developed. In addition, the researchers spent 2 days trailing through private galleries, souvenir retail outlets, cultural and Indigenous infrastructure and tourism information distribution points to count infrastructure, examine the range of products for sale and audit public (in particular, tourism) information on the local arts sector.

A 21-page, self-completion questionnaire was posted to artists, arts businesses and cultural workers listed on the database and questionnaires were also hand delivered by the Regional Arts Development Officer, James Giddey. Forty-five completed questionnaires were returned. The survey took approximately 1 hour to complete and, while it is very likely that some people would have had difficulty filling it in, the response rate was positively affected by the survey being distributed three weeks after the 2006 Census of Population and Housing, another complex instrument.

To enable national comparisons, creative practice categories and questions on income, employment, training, funding and career highlights were taken from Throsby and Hollister's 2003 survey of Australian professional artists. Cathy Henkel's questionnaire for her 2006 analysis of the screen industries in the NSW Northern Rivers region was also referenced. Follow-up interviews, strategic conversations and focus groups were conducted with 41 people.

Cultural infrastructure in Broken Hill

From the 'Roaring days' of mining there are a large number of local hobbyist and amateur cultural organisations, some dating back more than 50 years, that now have an ageing and declining membership. They include the Willyama Art Society, the Broken Hill Philharmonic Society, the Repertory Society, the Inland China Painters, Poets in the Pub, the Cameron Pipe Band and the Barrier Industrial Union Band. The 'professional' cultural organisations—publicly funded to (amongst other things) act for development and create the potential for new activity—include Broken Hill Regional Gallery, West Darling Arts, Broken Hill Art Exchange, Broken Hill Library and the Writers Centre.

The Indigenous arts sector has a low profile and is surprisingly—given high numbers of international tourists—underdeveloped. However, there is a recognised Barkindji art 'style' from the Central Darling region associated with the work of Badger Bates, Phillip Bates and Murray Butcher (Gibson 2008) and West Darling Arts has a specific focus on developing Indigenous arts and crafts. In 2008 the Broken Hill Regional Art Gallery introduced the Far West Emerging Aboriginal and Torres Strait Islander Art Prize to encourage emerging artists within the region.

The first private art gallery in Broken Hill was opened in 1972 and there are 25–30 private galleries in Broken Hill and Silverton. The galleries range in character from retail outlets with regular opening hours showcasing a number of local and other artists, such as the Silver City Mint and the Horizon Galleries; to smaller galleries in private houses around suburban Broken Hill focusing on the work of one or two artists, such as Boris Hlavica's Photographic Gallery; to showrooms in artist studios, such as Deidre Edwards' studio and framing workshop; to unique spaces such as Thankakali Gallery, in the cellars of an old brewery building, and the John Dynon Gallery in an 'authentic outback dwelling' in Silverton.

The so-called 'ghost town' of Silverton, 25 km north-west of Broken Hill, is an important local centre for the production and sale of art. The current tiny population of around 50 includes a number of visual artists and four galleries selling locally created art and craft. Silverton has also been a vital component of the region's attractiveness and development as both a film location and tourism destination. The 'magic light', the high number of clear filming days, the desert landscape and the industrial infrastructure are also reasons for the region's success as a film location. Since the 1960s more than 20 feature films and 200 commercials have been filmed locally. *Mad Max 2, the Road Warrior* (1981) and *The Adventures of Priscilla, Queen of the Desert* (1994) are the best known and are heavily promoted in tourism literature. As a screen worker commented: '*Mad Max 2* was filmed here 25 years ago and people are still coming to find it.' Film crews also praise the 'can do' attitude of the locals and the fact that they tend to leave film visitors alone 'to get on with it'. One screen worker remembered that:

> Mel Gibson said, when he was out here filming *Mad Max 2*, that there were 30 to 40 people coming out to watch the car stunts, but they never approached him for an autograph, they just sat and watched the cars rolling. On one of the last days of filming he actually went over to this group and said 'You've been watching it all?' and they were like, 'Yeah, gidday Mel.' 'How are you?' 'Love the movie!' and then one boy stood up and asked him for an autograph. Five weeks of filming: one autograph.

The community may not be 'rubbernecking' but are excited by the film work and are helpful to productions. As another screen worker said: 'When there is a film crew in town there is a buzz about and everyone in Broken Hill wants to be an extra.'

Another unique cultural and tourist location is the Sculpture Symposium on Sundown Hill in The Living Desert Reserve where, in 1993, visiting and local sculptors worked with sandstone from nearby Wilcannia to create their sculptures.

Who are the artists of the Far West?

Visual artists, craft practitioners and designers make up more than half (51 per cent) of professional creative workers in the region. Other occupations include musicians, screen workers, writers, community-based artists and cultural managers. Two-thirds of practitioners work in specialist areas, including basket maker, photographer, cartoonist, casting director, circus trainer, fractal imaging, graphic designer, sculptor, jeweller, short story writer, wood worker, restorer and stockwhip maker. Thirty eight per cent are 'established' practitioners who have been working in their creative occupation for longer than 21 years and 31 per cent are 'emerging' and have been practising 5 years or less (see Figure 1).

The two groups have different professional and business development needs, with established practitioners more likely to need support using new technologies for business efficiency and the emerging group more focused on developing practice skills and establishing a business. Emerging and mid-career artists expressed a desire to learn from successful older artists—'they have a lot to teach us'. Seventy-one per cent of artists are older than 45—with 40 per cent older than 55—and qualitative research has shown that even artists who move to the region are often older, more established artists with existing markets who relocate for both economic and lifestyle reasons. One interviewee remembered that:

> I fell in love with Broken Hill, with the desert, when I was out here shooting a car ad, then again for a [magazine] shoot ... then I saw an ad in the paper so I came out and now I've got a house and everything.

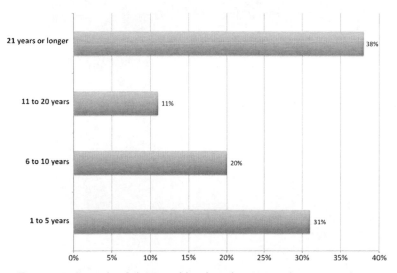

FIGURE 1. Length of time working in primary creative occupation.

Another said that:

> I brought my market with me. I send my work to established outlets elsewhere ... I have a network of galleries around the country ... And I don't think that artists do actually retire. You paint 'til your hands give out.

Only 18 per cent of artists were younger than 35 and only one artist was younger than 25. The qualitative research told a story of younger artists leaving the region for Adelaide or Sydney to access professional skills, markets and 'more exciting' lifestyles.

Artists were asked to break down the time they spent working (see Figure 2) and to itemise their annual income from July 2005 to June 2006 into money received from (1) their creative practice, (2) arts-related income (such as teaching and arts administration), and (3) income from work not connected with the arts. The average earned annual income from all work was $44 331—with the median $45 000. This was higher than the average earned income in Broken Hill of $36 834 (ABS 2008). Average annual income from creative work alone was $22 497; which compares favourably with the national figure for professional artists where the average income from creative work is $17 000 (Throsby & Hollister 2003).

Only 25 per cent of creative makers had specialist arts qualifications and, at the time of the survey, only 11 per cent were undertaking any form of training, either formal or informal. Overall, the results indicate lower local levels of formal and specialist qualifications than found in the national profile of artists (see Throsby & Hollister 2003). TAFE certificates are the most common formal qualifications. In keeping with a tradition established by the Brushmen of the Bush, 62 per cent of artists described themselves as 'self-taught' and half had learned 'on the job'. Nearly half had also accumulated skills through workshops or short courses,

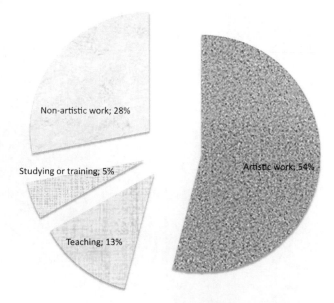

FIGURE 2. Division of time spent working, July 2005 to June 2006.

38 per cent had received training or work experience through a private teacher or practising professional and 13 per cent had been mentored. Not surprisingly, when asked to specify additional training needs the most common response was a desire to participate in arts exchanges with successful artists, both local and from 'Away'.

Local practitioners also have very low levels of membership of professional organisations and associations. Eleven per cent were members of the National Association for the Visual Arts (NAVA) and 13 per cent were members of local organisations, including the Willyama Art Society and Broken Hill Regional Art Gallery. Focus groups reiterated a general lack of knowledge about the professional organisations and specialist arts support agencies of metropolitan Australia.

The business of art

Most artists in Broken Hill are sole traders (36 per cent) or in a partnership arrangement (21 per cent). Eleven per cent are companies and 16 per cent are employed by someone else's business. Other business structures used are incorporated associations and family trusts. The 26 small businesses that participated in the survey currently employ a total of 77 people, including owners, full-time and part-time staff; an average of three per business. Less than half (42 per cent) are being guided by a business plan in their creative business and only 41 per cent have ever received professional business advice.

The average annual marketing spend was $3818, or 17 per cent of total income earned from creative practice. While 77 per cent of artists are primarily responsible for promoting their creative work and developing markets, only 31 per cent of respondents have a marketing plan in place. The main venues or outlets used to display or sell work services are, in order: private galleries and venues, repeat customers, commissions and through exhibitions or performances in public venues. While most (but not all) artists use computers and the Internet in research, creative work and office and administration work, only 36 per cent used computers for recording and tracking clients or artworks on consignment and only 11 per cent used e-commerce.

Arts businesses are aware that they need to spend more time and resources on building markets and outlets for their work but, while most understand the local market very well, the distance—both geographical and psychological—to 'outside markets' is vast. They also see that they need to share knowledge more often and cooperate to build audiences outside the region—so they are less dependent on seasonal tourist sales—and to promote the brand of Broken Hill as an 'arts place'. But factionalism, long-standing divisions, and professional rivalries hinder co-operative effort.

Creative businesses are optimistic about future prospects; 36 per cent described their businesses as 'growing' over the next 3 years, 38 per cent as 'sustainable', 17 per cent as 'commencing', with only 9 per cent describing their businesses 'in decline'.

Arts markets

Local consumption is the largest part of the market, with local residents responsible for 39 per cent of total sales and tourists accounting for 30 per cent (see Figure 3).

There are a number of serious local collectors of Broken Hill art and established visual artists observed there is more interest from this group when an artist is 'starting up', or that the collectors 'keep an eye out to get in on the beginning of something—when you start something new'. However, local purchases were characterised as: for family presents and interior decoration; gifts from work colleagues for someone leaving town (such as a mine worker, teacher or 'visiting medico'); and people who are leaving town purchasing a 'memento' of their time in the Far West (again, usually professional mining, education and health workers). One artist said: 'When they [people who have been working in the region] do the big house sell-off when they leave town, they go around and buy up what they like, and they spend big—it's a keepsake.' A well-established visual artist commented that he had developed two distinct styles of work: one that fitted the interior design market and the other for commissioned works, commonly paintings of mines bought by or for people who had been working in that sector.

The tourism market is seasonal and local buyers sustain the artists throughout the low and shoulder tourist seasons. As one artist described this situation: 'December to March are pasta and salad days, but when April comes, with the tourists, it means steak for dinner.' Artists also talked about a price ceiling for local sales, which one artist described as 'around $3,500, which is difficult to get above. The same painting if sold in Sydney would fetch a price of $9,000.' Visitors with more outwardly visible discretionary income are (not surprisingly) more likely to purchase 'big ticket' artworks. One artist talked about, 'the Subaru Forrester drivers, who, when you see them pull into the car park, you know will leave with something from our gallery'. Other likely purchasers of outback artworks are visitors on long tours who use Broken Hill as an entry and exit point for travels in the Outback and buy artwork as a memento of their trip on their return journey.

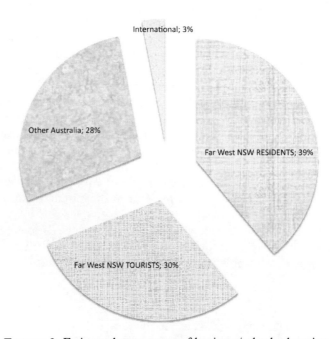

FIGURE 3. Estimated percentage of business/sales by location.

The USA and Europe account for most international sales. Sales to international visitors need to be 'crated and freighted', which affects the sale price. 'When you box it up and give them a price', one artists commented, 'they go "oh my god!"— TNT is like $150 for a painting'.

Life in the West—place and practice

One-third of creative workers were born in the Far West and 20 per cent have lived in the area for more than 20 years. Of the locally born artists, a number had moved away for many years to travel and establish careers, and then returned. One interviewee spoke about being asked why they had moved back to Broken Hill:

> Not all of them [local people] understand it. They're usually from the mines, and their idea of the place has been entrenched because of their work. They don't see the artistic side. Also, some people live here and don't get the opportunity to see different places and experience other things, to compare. I was born here and then moved back, so I'm an A grouper. My friends, a couple of them were artists from Broken Hill who went out to explore the world, then came back for the isolation and stuff.

Twenty-seven per cent of artists had moved to the area in the past 5 years. 'Five years', said one:

> that's sort of the milestone in Broken Hill. I've seen lots of very talented and productive artists move away. Could be many reasons. Maybe people aren't ready for their work, like they're not mainstream or landscape artists ... I think they should stay because new stuff brings change ... I dunno, maybe they're sick of banging their heads against the wall.

Many had previously visited and 'fallen in love' with the landscape and/or the 'magic light' and/or the sense of 'freedom', but had made the decision to move to Broken Hill to have more time to spend on their arts practice and because housing and studio spaces are 'cheap'. During a focus group one artist said:

> I was going to live in Spain, but we had to wait two months until our daughter was born so I thought I'd travel around Australia. And I came across Broken Hill three times in all that. It was the friendliness, the lighting, the colours. Just the whole environment. It was almost like a magnet, bringing me back here.

Another commented: 'I moved here to get on with artistic endeavours in my life but I've taken a real liking to it ... I had been here before ... [and] I wanted the space and community, the freedom.' Yet another said that in Broken Hill they could have: 'A house for ourselves, for $50 000, out the back of a gallery. I have my own gallery and four bedroom house out the back of it. It's perfect'. Another sub-group are people 'escaping' some trouble in their life who see Broken Hill as a 'fresh start'.

Figure 4 shows the top-ranked disadvantages for creative making in the Far West. Alongside feelings of isolation and lack of local resources there was also a sense of the insulation of the local community and the lack of 'cultural stimulation',

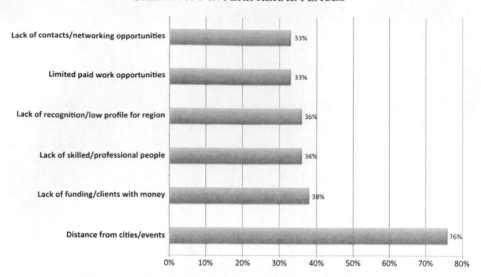

FIGURE 4. Top six disadvantages of living in the Far West.

including a lack of access to cultural events, fellow artists and being able to 'keep up with what is going on in the cities'. Twenty-four per cent of artists also felt that the region placed 'too much emphasis on landscape art'. The cost of materials and the availability of local suppliers was also a concern. (A somewhat despondent artist wrote across the questionnaire, 'Unfortunately I had only five choices, otherwise I would have probably chosen nearly all of them.') However, the vast majority of artists (84 per cent) felt that living outside of a capital city had had a mainly positive impact on their practice. The inspiration of the landscape and the visual environment were the most important positives for creative practice followed by the lower cost of living (see Figure 5). As a sub-group, traditional and craft practitioners talked about the access to local, raw materials and the 'genuine' quality of their work. Other advantages were: enhancement of personal health (31 per cent), less crime (24 per cent) and less competition (18 per cent). A large

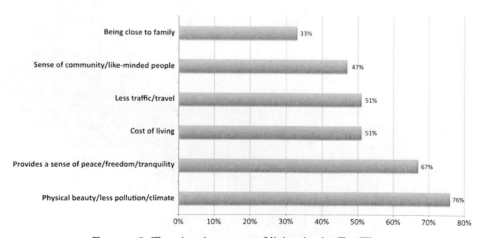

FIGURE 5. Top six advantages of living in the Far West.

82

number of artists contrasted the 'freedom' of their life in the West with their perception of the 'stress' of a Sydney-based artist's lifestyle:

> I spent three months recently in Sydney and I didn't pick up a pencil or a brush the whole time. I was too busy and stressed. I have more choices here, to relax more and paint. In Sydney there's a pressure to have a different lifestyle. (Interview, 2007)

Similarly, another artist commented:

> This is what I sense from people in Sydney, I mean, they talk about the brightness of our colours and sculptures. Whereas Sydney I find very oppressive. I think there's a real sense of creativity in this community, a kind of feeling or vibe.

Artists who had moved to the region described city-based pressures on creative practice to conform to art 'fads and fashions' (see also Gibson *et al.*, this issue). As one artist said of his life in Sydney: 'I used to feel as though I had to conform to certain styles, like my creativity was being squashed in a box.'

The desert and mining landscapes and the quality of local light exerted a powerful influence on every artist I spoke to and a number of them are active in environmental causes. 'It's very uplifting', one commented,

> I find as I'm driving home to Broken Hill I can feel it. You leave all that stress behind and get a new perspective. I find it very stimulating. I think there's a physical thing too. Of clarity. Like the view from the window. You look out and everything's so clear. I've been to many other countries ... and I haven't found a clearer view. The vivid blue sky, the sharpness of the landscape, and the variety.

Artists who had painted the outback for many years described a deep psychological connection with the desert landscape. One artist said that, after years of painting the drought, she dreams of deep underground pools of water and had even developed a new painting style around deep blues and greens—'not that there is a market for that work out here. People come here to buy Outback art'. One artist wrote that 'Art making in context of desert rather than more lush conditions provides a timeless context.'

Conclusion

The Broken Hill research reveals that life on the periphery is both enabling and disabling for artists. The Arid Arts and Crafts community of Broken Hill—as well as the Santa Fe Art Colony, the Småland glass designers of southern Sweden, and the Blue Ridge Mountain craft makers of North Carolina—are good examples of how creative practitioners can use their 'remoteness', wild landscapes, and local culture, traditions and materials in making and marketing unique creative products that successfully tap into consumer's imaginations. But more research is needed to develop a better understanding of creative practice in remote communities, the impacts of peripherality on the individual makers who are the source of wealth for this sector, and the opportunities of the evolving information economy. In Broken Hill, 'remoteness' means limited types of creative making; wariness of newcomers and new ideas; the loss of young people; limited access to business expertise, production services and training; lack of cultural stimulation; and high transport costs. 'Isolation' means a unique local landscape and culture; a friendly community; lower-than-city costs for accommodation and studio space; freedom from city-based art 'fads', stress and busyness; and a 'quality of light', time and a clear view.

Acknowledgements

This article is based on my research for the Far Western Development Board and West Darling Arts (Andersen and Andrew 2007) with funding from the NSW Department of State and Regional Development, Regional Arts NSW and the Australia Council for the Arts. Many thanks to my co-researcher, Jane Andrew. I acknowledge the ideas generated by numerous A-groupers, B-groupers, and people from Away who contributed to the research—especially Eileen Braybrook, Ellenor Day, Geoff de Main, Fiona Ellis, James Giddey, Cathy Henkel, Kathy Kennewell, Elizabeth Rogers, Robert Sidford, Grant Smith, Bronwen Standley-Woodroffe, David Throsby and the members of the Research Advisory Group—and thank them for their generosity.

REFERENCES

ALLAN, J. (2008) 'Mining's relocation culture—implications for family, community and industry', Discussion Paper 1, Centre for Inland Health, Charles Sturt University, Wagga Wagga.

ANDERSEN, L. & ANDREW, J. (2007) *Quality of light, quality of life: professional artists and cultural industries in and around Broken Hill*, Far Western Regional Development Board, Broken Hill.

AUSTRALIAN BUREAU OF STATISTICS (ABS) (2006) 'Arts and cultural heritage in Australia: key issues for an information development plan', discussion paper, Australian Bureau of Statistics, Canberra.

AUSTRALIAN BUREAU OF STATISTICS (ABS) (2008) *National regional profile, Broken Hill, 2002 to 2006*, Catalogue No. 1379.0.55.001, Australian Bureau of Statistics, Canberra.

BROKEN HILL CITY COUNCIL (2004) *Cultural plan 2005–2010*.

BROKEN HILL HISTORICAL SOCIETY. A visit to Broken Hill in 1888: letter from Edmund Harral. Pages and photos from the past of Broken Hill, available from: http://www.geocities.com/bhhsi/ (accessed 18 October 2009).

ELLEM, B. & SHIELDS, J. (2000) 'Making a "union town": class, gender and consumption in inter-war Broken Hill', *Labour History* 78, pp. 116–40.

FAR WESTERN REGIONAL DEVELOPMENT BOARD (2003) *Strategic plan 2004–2010*, Broken Hill.

GIBSON, L. (2008) ' "We don't do dots—ours is lines": asserting a Barkindji style', *Oceania* 8(3), pp. 280–98.

GREER, G. (2006) 'Australia's Lowry is finally being recognised by its artistic elite—but is it for the right reasons?', *The Guardian* 8 May.

HENKEL, C. (2006) *Imagining the future 2: screen and creative industries in the Northern Rivers region*, Northern Rivers Regional Development Board, Lismore.

LEA, T., LUCKMAN, S., GIBSON, C., FITZPATRICK, D., BRENNAN-HORLEY, C., WILLOUGHBY-SMITH, J. & HUGHES, K. (2009) *Creative Tropical City: mapping Darwin's creative industries*, Charles Darwin University, Darwin.

THROSBY, D. & HOLLISTER, V. (2003) *Don't give up your day job: an economic study of professional artists in Australia*, Australia Council for the Arts, Sydney.

TOURISM RESEARCH AUSTRALIA (2008) *Tourism profiles for Local Government Areas in regional Australia, City of Broken Hill*. Tourism Australia, Canberra.

Creating an Authentic Tourist Site? The Australian Standing Stones, Glen Innes

JOHN CONNELL & BARBARA RUGENDYKE, *University of Sydney, New South Wales, Australia; University of New England, Armidale, Australia*

ABSTRACT *Developing and sustaining a tourist economy in regional Australia has required innovative strategies. Shifts towards cultural tourism have resulted in the revitalisation of heritage and the development of tourist sites that are authentic (re)presentations of past landscapes and peoples. In Australia, where European heritage is comparatively young, lateral thinking, creative licence and municipal efforts have been required. The Australian Standing Stones at Glen Innes were constructed to enhance the region's perceived Celtic heritage and stimulate tourism. Tourists at the site have discerned elements of heritage and authenticity despite their recent construction. Creating a distinctive tradition, however tenuously linked to history, can be an effective means of branding place and stimulating tourism.*

> Stone sentinels
> Spokes of a wheel
> Silent and strong
> Contagious feel
>
> Exacting, compelling
> Hearts miss a beat
> Majestic, enthralling
> A delight to meet
>
> (Mathew n.d., p. 7)

As cultural tourism has universally become of greater significance, tourism authorities have sought to stress elements of local and regional heritage by a variety of means in order to emphasise authentic connections between present and past landscapes and peoples, and so stimulate tourism. The past has been re-imagined and transformed through new discourses, alongside various forms of creativity as places have sought to differentiate and brand themselves in a crowded tourist marketplace. This paper examines how one part of northern New South Wales has not just reinvented itself through a particular historic theme but creatively constructed a major new tourism facility to sustain and monopolise these claims, and how this has impacted on tourism and urban identity.

For almost two decades the Glen Innes region in northern New South Wales has proclaimed itself to be Celtic Country and the town has hosted the annual Australian Celtic Festival since 1992. At the southern entrance to the town, on the main highway, a large stone monument proclaims a welcome to 'Celtic Country' (see Plate 1). The main street is marked at regular intervals with banners proclaiming Celtic Country, illustrated with the stone monument, now an icon of the town. The Town Hall flies the Australian and Aboriginal flags alongside those of the Isle of Man, Scotland, Brittany, Cornwall, Ireland and Wales. Every week on Friday at noon a lone piper plays 15 minutes of traditional Scottish airs from the balcony of the Glen Innes Town Hall. The cover of the principal publication of the local tourist authority (see Plate 2) is dominated by Celtic references and a kilted Celtic bandleader. The back cover (see Plate 3) portrays Celtic bagpipers in front of the Standing Stones, and similar pictures and references dominate the content of the booklet. The Standing Stones have their own distinct brochure (see Plate 4) within which there are hints of a glorious Celtic past in references to the battle of Culloden and the photograph of a mediaeval warrior in armour.

The basis for this Celtic link lies in Glen Innes, a town of around 6000 people, having been settled 'largely by Scots in 1838',[1] though there is no indication of what proportion of all settlers they were or how many of them came from the Celtic fringes of Scotland. One CD of local music, available in the tourist office, claims that Glen Innes was originally settled by Scottish pastoralists and Cornish tin miners around 1850. The Ben Lomond Mountains and the tiny settlement of Glencoe lie to the south, as also does the Stonehenge Road. Numerous other areas of regional Australia, including Inverell not far to the west, have similar claims to Celtic connections and any primacy of Glen Innes cannot be demonstrated. While other parts of Australia had pioneer settlers from peripheral parts of Britain, few

PLATE 1. Southern entrance to Glen Innes.

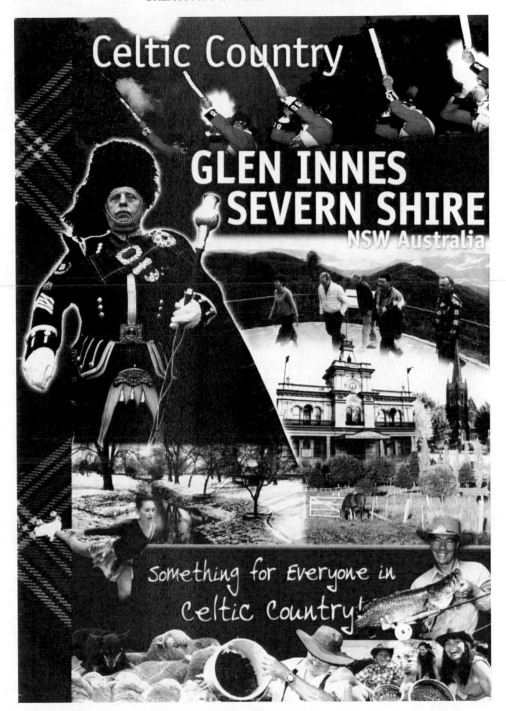

PLATE 2. Celtic Country tourism brochure.

have sought to proclaim this link in tourism promotions. Glen Innes has thus managed to claim a unique place in Australian tourism, with its name (and regional names) being a key link to distinctiveness.

PLATE 3. Celtic Country back cover.

While heritage tourism has sought to emphasise and market authenticity in terms of the continuity of visual images (and often performances) with the past, there have been frequent debates over the extent to which authenticity can be sustained, rediscovered and restored. At Glen Innes the construction of the Standing Stones,

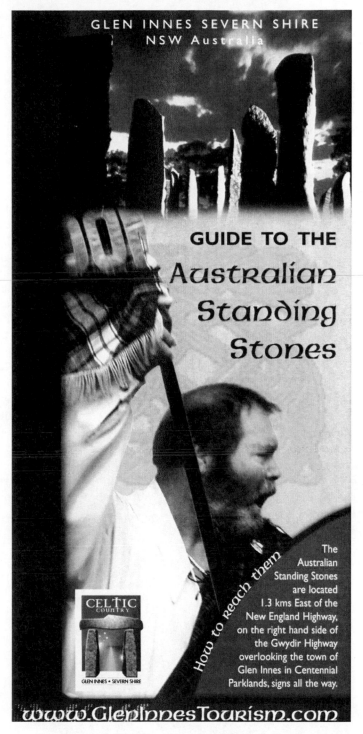

PLATE 4. Standing Stones tourism brochure.

as a key plank of the town's identity and the principal tourist site, has seemingly taken this one step further by creating a new landscape intentionally replicating, and reminiscent of, idealised Celtic landscapes in another hemisphere.

The Australian Standing Stones

The Australian Standing Stones, on a hillside on the edge of, and overlooking, the town (see Plate 5), were constructed in the early 1990s and officially opened by the Governor of New South Wales in 1992, 'In recognition of the involvement of the Celtic Races in the building of the Australian Nation.' They are perceived as 'unique in the southern hemisphere' and 'probably the first of their kind built anywhere in the world for more than 3500 years'. At the same time they have been distinguished, notably in earlier tourist publications, by being contemporary stones with no links to a pagan era, having been 'built in Christian times, in the predominantly Christian community of Glen Innes' (quoted in Auster 1996, p. 114). Such statements allayed some local concerns that the stones would lead to pagan ceremonies and, in the brochure published to mark the opening of the Stones, the Chairman of the Glen Innes Celtic Foundation noted: 'it had never been intended and never will be, that the object of the Standing Stones project is to conduct heathen rites. In fact the four stones marking the cardinal points of the compass and the Southern Cross also represent the cross on which Christ died for all of us'. It was neither pagan nor a transplant from the northern hemisphere.

The project began through the work of a 'small dedicated group of citizens who wanted to mark Glen Innes' Scottish heritage'. A key member of the group was the town's tourist officer. It was stimulated by the decision of the Celtic Council of Australia in Australia's Bicentenary Year in 1988 that there should be a national monument 'to honour all Celtic peoples who had helped pioneer Australia', and Glen Innes won a national competition to host the Stones. Both the town and the Celtic Council perceived the Stones as a 'focal point for local tours of Japanese and other tourists' (Letter from the Celtic Council of Australia to the Glen Innes Town Clerk, 24 July 1990).

PLATE 5. The Australian Standing Stones.

The main feature of the Standing Stones is the circle of 24 stones, representing the 24 hours of the day. Outside that circle four stones mark east, west, north and south. These four stones, with a single stone inside the circle, form the Southern Cross, symbolising the link between the old and new worlds. Three stones in the centre of the array represent a northern stone (for the Gaelic-speaking Celts of Ireland, Scotland and the Isle of Man), a southern stone (for the Brythonic-speaking Celts of Wales, Cornwall and Brittany) and a middle stone, representing all Australians. Outside the circle there are also two other distinctive stones: the Gorsedd Stone, for the Cornish and Welsh, and the Ogham Stone, for the Irish. The Ogham stone has an inscription carved in Ogham, the oldest writing known to have been used by the Celts around the fifth century, which translates in Gaelic as Gleann Maqi Aongusa, the Glen of the Sons of Angus, or, more simply, Glen Innes. The inspiration is said to have come either from the Stones of Callanish on the Isle of Lewis in the Hebrides, or the Ring of Brodgar in the Orkney Islands. Each of the stones weighs about 17 tonnes, and comes from the local region. Indeed, the Celtic Council gave primacy to Glen Innes for the site because of its access to appropriate granite. The stones were rough hewn, to accord more with Celtic imagery, and sponsored by various organisations ranging from several clans, such as the Clan Cameron, to the Highland Society of New South Wales, Knox Grammar School (Sydney) and British Airways.

Immediately south of the assembly is a viewing platform on Tynwald Hill, named after the Manx parliament, the longest continually serving parliament in the world. On the lower slopes of Tynwald Hill is a rock in which is embedded the sword of Excalibur, the mythical sword of the probably equally mythical King Arthur, that is a part of Celtic mythology from Brittany through Glastonbury to Wales. The sword stone was dedicated by Roy Reeve, the British Consul-General, in May 1994 to 'the Welsh communities' of Australia.

On a different part of the hill is a 'Wall of History' that is being created with stones taken from historical sites from 'all over the Celtic world'. Various stones, mostly about the size of a cricket ball, are embedded in the wall (see Plate 6) and come from either particular rock formations in such places as Tintagel and Truro in Cornwall, the Isle of Jura, from Celtic stone sites such as the Ring of Brodgar, in the

PLATE 6. The Wall of History, Glen Innes.

Orkneys, and from castles such as Ferniehurst and Dunnottar in Scotland. One is Caithness stone from the birthplace of Major Archibald Clune Innes, reputedly the earliest colonial landowner of Glen Innes. Some stones have been presented, such as that from Glen Maye beach, Isle of Man, dedicated to the 'Manx people of New South Wales', or donated, such as that from Fleshwick Bay, Isle of Man, from the Queensland Manx Society, and even that from the Clan Johnstone in New Zealand: 'this stone of volcanic origin acknowledges our Glen Innes ties'. Every Celtic region and diverse symbols were thus given a place.

The Tea and Souvenir Shop is fashioned as a Crofter's Cottage, built of basalt and with a thatched roof, intended to be a replica of a Taigh Dubh, the small 'black house' of early Celtic peoples. It was inspired by a photograph of such a cottage that survived the battle of Culloden in northern Scotland in 1746. The Cottage offers 'Celtic fare and modern café cuisine'; the former include Cornish pasties, Dundee Cakes and Guinness, but chai latte tea is also available. Inside the cottage is a plaque inscribed 'This plaque commemorates the contribution of Scottish Gaels who since the days of Celtic speaking [sic] Governor Lachlan Macquarie have helped mould Australia (Council for Scottish Gaelic, January 26th 1988)'. The Cottage flies a Celtic flag and plays background Celtic music, while a stuffed, antlered stag's head, its neck adorned in tartan, majestically surveys the tea room.

Souvenirs for sale include Celtic herbal elixirs, to be taken 'with a wee dram', local honey and jams with tartan material covers, model sheep in tartan jackets, ponchos in Glen Innes tartan and tea-towels variously saying Scotland, Wales, etc., though one entitled 'Aussie beer and damper' was subtitled Glen Innes Celtic Country. Glen Innes has created its own tartan, in which many products are adorned, in five colours: light blue ('the clear day time skies of New England'), royal blue ('sapphires for which New England is renowned'), dark blue ('night skies in which the Southern Cross is predominant'), Red ('our Celtic (Scottish, Irish, Welsh, Cornish) blood links') and white ('fidelity with the Celtic past and traditions'). In the town the Tourist Office echoes the same themes and sells products which, other than sapphires and a handful of tourist knickknacks, such as fridge magnets, focus almost exclusively on Celtic themes. These include Celtic pottery (made locally), T-shirts and hats, 'highland preserves', spirtles (Scottish porridge stirrers) and three locally produced CDs, two from a local trio, Spirit of the Glen, and one from the Glen Innes Arts Council Choir. They include standards from the British Isles, including the Manx National Anthem, Annie Laurie, Land of My Fathers, but also the Standing Stones and Cootamundra Wattle.

Although viewing the Stones is primarily a passive experience it is possible to book a 'Celtic Crone Tour' which provides 'an exciting costumed guided tour highlighting ancient Celtic history'. The brochure further notes that 'Strange myths and legends surround the stones in Europe and while no such claims are made for the Australian Standing Stones, some visitors have felt a powerful, spiritual, influence as they walked through the array.'

The original intention of the related Australian Celtic Festival was to revive Celtic heritage and culture through song and language; as the Festival Chairman, Howard Eastwood, has stated:

> If you have Celtic origins you'll certainly be emotionally stimulated and
> spiritually inspired. Even those who don't will find the festival a

fascinating, educational and entertaining event. (Quoted in *Armidale Express* 19 April 2006)

Each Celtic 'country' is given equal time but particular 'nations' are singled out each year; in 2006 it was Bretons and also Galicians from Spain, as the Celtic world expanded further. The ninth Roman Legion and 42nd Royal Highland Regiment brought a different emphasis. The Festival traditionally begins at the Standing Stones with a skirl of pipes from a lone piper at dawn. The Festival involves Celtic music, sheep dog trials, pipe bands and the 'kirking of the tartan' and a multiplicity of related activities. Glen Innes also hosts the annual Festival of the Beardies, a festival much like many in rural Australia, but with its central theme around prizes for particularly distinctive beards in different categories. A key element of this festival, and other local activities, is Scottish pipe bands from the town and other nearby towns. There is a nearby Take a Wee Brek Fossicking Park and the town is developing events like Gourmet in the Glen, which began as Grapes on the Grass, a show for New England wine producers in 2004, but like other activities acquired a more Celtic name. Marriage celebrants, dressed in tartan, are also available, and the new Celticity has even inspired a pamphlet of doggerel verse (Mathew n.d.). Early plans for the division of the town into six different Celtic suburbs, and the construction of a Celtic cross, never eventuated.

Glen Innes thus widely promotes itself as a distinctive Celtic place, in every tourism context and, for example, at the annual Country Week Expos in Sydney, where various towns in regional New South Wales promote themselves to encourage the migration of urban residents to rural and regional centres. One man, dressed in kilt and sheepskin, spruiked its attractions, and tea bags with tartan tassels and biscuits with tartan labels were part of the Glen Innes 'showbag' handed out to visitors.

In a variety of connected ways Glen Innes has differentiated itself from other parts of regional Australia by proclaiming its Celtic heritage. However, it has done this primarily through the Standing Stones that have no continuity with the past, other than through mythology. As the Deputy Mayor of Glen Innes has said: 'The branding started with the development of the Standing Stones' [which has proved to be] 'incredibly useful after initial cynicism and scepticism locally [by] locals who thought it a bit corny' (pers. comm. 2007). One of those, a Glen Innes Councillor, has said: 'I was initially cynical "Let's do Stonehenge again and again" but it has acquired gravitas' (pers. comm. 2008). Glen Innes has proved more accepting of its Celtic turn than small North American towns where more evident 'fantasy themes' have been adopted to stimulate tourism, and creativity has brought division and resistance (Paradis 2002, 2004; Mair 2009).

The Stones are not so much the continuity of heritage but the reinvention of tradition (Hobsbawm 1983) in a visual form that is central to tourism. Much of the local tourist literature hints at authenticity, uniqueness and significance, but the presence of 'Australian' stones marking the Southern Cross emphasises that the site is quite different from those of Britain. Overall, it represents a dynamic claim to 'strategic inauthenticity': a deliberate re-creation of a sanitised or glorified version of the past, whether invented, borrowed or preserved (Gibson & Connell 2005), because of its commercial potential.

Whether Glen Innes' claim to Celtic heritage is widely known beyond the region is doubtful. Glen Innes is not usually a national destination for tourists but has tended to attract short-term visitors through being on a significant highway. Other than having the Stones, and declaring an affinity with and a history of Celtic life, Glen Innes is not very different from many other Australian country towns and hence the Stones have become both a site for passing tourist buses to have a break (close to the adjacent tea rooms) and a site for tourist groups to break journeys.

Tourists at the Stones

Auster has argued that there are three groups of tourists at the Stones: a 'first level' of those who have a break there, visit the toilets, enjoy the pleasant site and carry on, a 'second level' for whom being Celtic is a 'personal identity thing', who visit the site 'in a spirit more or less of pilgrimage' that may be both lighthearted and solemn, and a 'third level' 'where a few people are moved, in a way not necessarily related to Celticism as such, by a feeling of communion with something larger than themselves, in space and time' (1996, p. 113). For this last group the Stones may be 'a place of psychic integration, a place of folk memory' (1996, p. 115). While Auster indicates that only a 'few' people are deeply moved by the experience, he provides no sense of numbers and his differentiations are no more than suggestions, unrelated to survey data.

Many tourists have voluntarily entered their comments on the Stones in the Visitors' Book (though this provides little space for eloquence, hence comments often tend to be no more than a couple of words). Most are positive, as might be expected, and represent only a very small proportion of all visitors. Almost all those who have written in the Visitors' Books were from outside the Glen Innes region, usually from elsewhere in eastern Australia, but one resident of Glen Innes had written 'never get tired of this place'. Based on comments written mainly over a 12-month period in 2005–06 their perceptions indicate that many tourists, but no doubt those most likely to write in the book, derived some degree of inspiration from the Stones. Firstly, many found the site of some spiritual or mystical significance: 'touches the soul', ' a mystic place', 'energising', 'awesome, inspiring', 'very interesting energy here', 'mystic', 'spiritually uplifting', 'very spiritual—cool energy'. Despite the attempts to link the Stones with the Christian era, many such comments attest to 'new age' perceptions and consciousness.

Secondly, a substantial number of others, invariably of self-proclaimed Celtic (usually Scottish) heritage, found elements of nostalgia: 'with Scots and Irish heritage on both sides this place is a spiritual return', 'so far from Scotland but so close in spirit', 'brought back childhood memories', 'nostalgic', 'it stirs the Celtic heart', 'home from home', 'Celtic in Oz', 'Brilliant: proud to be of Welsh and English heritage'. Some transferred their comments into a simplified version of what might have been Scots: 'mc bloody mc beauty', 'long may yer lum reek [chimney smoke]' or fragments of Gaelic and Welsh 'Rath de ar an obair', though most could manage little more than 'och aye', 'och aye the noo', or 'begorra t'is good'. Others who were visitors from Scotland, Wales or Ireland found moments of nostalgia: 'making me homesick', 'a wee touch of hame', 'felt really at home', 'great to see a wee bit of Scotland here' and 'very Scottish'. One visitor from Nova Scotia, Canada pointed out 'Nova Scotia has Scots too'.

A third group simply found the site and stones larger than anticipated, though some comments were clearly tongue in cheek: 'get your rocks off', 'wikid', 'authentic sculptural work', and 'they are very big stones'. The third groups at least were not swayed by notions of Celtic heritage in either a general mystical form or through a link to their own pasts. One from Scotland noted: 'not the real thing'. While such 'non-believers' are probably a majority, they had at least been drawn to the site.

A survey of tourists in November 2009 indicated that many had arrived merely because their coach stopped there, or because it was 'another stop on the road' as they traversed eastern Australia. Significantly it was, however, a key stopping point and place to relax, rest and eat and drink or enjoy the sights. The Stones have become a distinctive place to take a break. Many such tourists have little interest in the Stones even though the Stones have brought them there. However, at least as many had made a special effort to stop at the Stones, often for a repeat visit.

Interviews with the tourists revealed similar categories of perception from those derived from the Visitors' Books. Some 39 interviews were conducted over 4 days, from Friday to Monday, encompassing a school holiday weekend.[2] Of these, seven were individual interviews, while the others involved various groups, hence 96 people participated in interviews. Eight were overseas tourists, 29 interstate visitors, 32 from elsewhere in NSW and 27 were local people. Their responses parallel and extend those in the books. Firstly, some found the site of significance, but proportionally fewer than in the Visitors' Books (emphasising that they are filled in by the more impressed visitors): 'one year I came here in the middle of winter and there was fog and I loved it', 'impressive; it was misty and we got a real feeling of ancient Britain, an atmosphere of mystery I would say'. Secondly, many were anxious to claim Celtic (usually Scottish) heritage and their own rationale for visiting the site: 'my grandfather was Scottish', 'I'm a quarter Scottish; my grandmother was Scottish' and souvenirs were brought for living relatives: 'I bought Mum a thistle sticker; she'll love that'. Ancestry could be complicated: 'The wife has Celtic blood, a mixture of Celtic and Moorish; you know the Welsh.' Some wanted more information: 'My grandparents were Celtic; I thought it would be good if they [the Stones] had something at the bottom of them so we can work out which one is our clan.' A third, but small, group refused to be taken in: 'Disneyland to be honest; when you've seen the original you know', 'new history isn't it; I know it's to do with settlers and all that but it's a bit manufactured', 'I've been to Scotland and the original is better', 'No comparison to the real ones; the real ones are more worn and old-looking and you're not allowed to touch it any more. Whereas these are more Australian, rugged looking.' That had not discouraged their general enjoyment of the experience. Some were merely puzzled: 'quite good but stuffed if I know what it's about'.

The surveys revealed three other ways of thinking about and experiencing the Stones. The majority of visitors simply enjoyed the Stones as an impressive site, with some vague historical connotation. 'I was very impressed with the stone from the thirteenth century battlefield', 'it's quite spectacular from up the hill', 'I love the old stone buildings—look at the roof—the logs and the thatched roof', 'I'm very impressed; for an area of this size what they've done is very impressive and that's without really understanding the whole Celtic thing—they had the foresight to develop it', 'I'm intrigued by the flat rock, how they found it and managed to lift it into place'. An American visitor observed: 'Very interesting; no idea they'd have something like this in Australia; I've been to Stonehenge, quite a project. Similar sort of thinking, similar principles.'

It was also a somewhat generic tourism place: 'We love them, we always enjoy them; we love coming here', 'We come up whenever we come to Glen Innes', 'We would definitely come again if we were in the vicinity and we would recommend it to others', 'I think we'll visit again now we know they are here', 'it's a nice drive for the day', 'The kids like the walls because they can walk on them'. It was as pleasant a place as many other tourist sites.

Finally, many local people visited the site because of its distinctiveness, the view from the Stones, the exercise in walking there or because the coffee at the Cottage was reputed to be 'the best in town'. While some claimed Celtic ancestry ('I really like them because I'm into the whole Celtic thing'), and others eulogised the Stones: 'they're fabulous, they're great; I've been here since they went in; they don't look as raw now, they're starting to blend in'. The majority, like many visitors, simply saw it as a pleasant neighbourhood site, to come to themselves or to take visitors: 'we bring all our visitors here. This'd be about the fiftieth time', 'we come up here a lot, we just bring the kids up and let them run around . . . we always come up when we have family visiting, and of course there is Crofters Cottage [where] they serve great coffee', 'it's nice up here in winter when it's covered in snow and the fog, you know the Scottish mist'. Many local and also distant visitors had come several times. The site has taken on a local resonance, matching the diversity of tourist experiences.

Conclusion: inventing heritage tourism

While the local tourist office necessarily proclaims the authenticity of the site, and the similarities between these stones and those in Celtic Europe, their recency largely denies this claim. Similarly, Glen Innes' claim to Celtic heritage has no great historical merit, even if largely unchallenged. The Standing Stones represent creativity as a form of strategic inauthenticity where an obscure historical past has been transformed and reified through their construction as the key symbol of the past. In a broader context, even Australian Celts 'are a fine example of what has come to be called the imagined community' (Auster 1996, p. 115). However slippery and elusive authenticity may be, the Standing Stones have minimal regional heritage, testifying primarily to the creativity of council officials with the support of the Celtic Council of Australia.

More specifically, even the location of the Stones challenges notions of Celtic history

> The reality in Glen Innes, is a hillside on the edge of a small town, with gum trees. The glaring Australian light, and the water tank overlooking the site, seem fit to banish any trace of Arthurian romance. Can Celticism flourish here? Apparently so. (Auster 1996, p. 113)

Effective place branding has ensured that it does so. Yet in many ways, through this branding, the Standing Stones are more obviously a low-key theme park: a themed Celtic environment with tangential claim on history. Indeed, the Standing Stones are emblematic of themed environments:

> socially constructed built environments . . . designed to serve as containers for commodified human interaction . . . material forms that are products of a cultural process aimed at investing constructed spaces with symbolic meaning and at conveying that meaning to inhabitants and users through symbolic motifs. (Gottdiener 2001, p. 5)

Conveying that meaning has been achieved through a 'presentist bias' (Lowenthal 1985, p. 362) that portrays and purifies the past in ways that suit present values.

In this way, despite the lack of authenticity, the Stones have created a distinct sense of place, and a landmark which has both stimulated tourism, with Glen Innes becoming a significant stop on the New England Highway, and enabled at least some of those who visit not merely to enjoy the break but to derive a deeper satisfaction from the Stones, as they stimulate both nostalgia and new age consciousness—postmodern forms of detachment. The annual Australian Celtic Festival further emphasises their critical place; a distinct festival, however tenuously related to place, as in other towns such as Parkes and especially Bundanoon (Brennan-Horley *et al.* 2007; Ruting & Li 2010), has further contributed to tourist development in Glen Innes. What may be a theme park for some, or an unusual toilet break for others, can be a symbolic experience for at least a small proportion of those who visit, and perhaps especially those of British ancestry who have kin or have seen stone circles on the other side of the world. Even while recognising the essential unreality of the Stones, the illusion of authenticity conjures up and stimulates images and memories.

Developing a Celtic theme has been beneficial beyond tourism, with the branding of distinctiveness placing Glen Innes within other contexts. At annual Country Week Expos in Sydney 'people got a positive perception of the place [and the] penetration of Celtic branding has been crucial when twenty or thirty country towns all try to stand out from each other' (Deputy Mayor, pers. comm. 2007). The Mayor's wife has said that Celtic themes have 'been the most effective branding exercise if you wanted to be cynical' (pers. comm. 2008). Whether through tourism or a more nebulous but positive sense of difference, creative branding has given Glen Innes more than a mere local presence.

Embracing difference and embellishing this with inauthenticity has boosted tourism. There are close resemblances between the transformation of Glen Innes and the globalisation of Irish pubs, similarly 'steeped in quasi-Celtic décor' (Patterson & Brown 2007, p. 46), parallel parodies of distant places and triumphs of marketing. Glen Innes almost literally has left no stone unturned in developing Celtic links. It has chosen a specific period of history of regional significance, and exulted in it. In doing that it has excluded other historical periods, along with other settlers of non-Celtic heritage. Unlike in many other Australian towns, little attention is given to Indigenous history: the Cooramah Cultural Centre is simply marked on one map but otherwise ignored in tourist literature. The town's Land of the Beardies Museum has a Celtic Room but no reference to Indigenous history. One component of history has largely been removed at the expense of another, more tenuous one. A veil has been placed over the past and present geography of Glen Innes, as 'the geography of differentiated tastes and cultures is turned into a pot-pourri of [Celtic] internationalism' (Harvey 1989, p. 89), with the creation of an artificial space and a dissonant heritage: a single core identity that brands Glen Innes as Celtic and erases differences, conflicts and alternative senses of place.

Deliberately creating a distinctive local tradition, however tenuous its historical connections, and however it mythologises and reifies one era, has proved an effective means of branding place and stimulating tourism. It clearly distinguishes Glen Innes from other regional centres, provides a memorable tourist site and generates income and employment for the town through the commodification of an imagined and constructed past. Clever marketing of 'Celticity' has placed

Glen Innes on a tourist map and contributed to regional dynamism. Ingenious creativity, invention and inauthenticity have enabled a vague history to be transformed into a new material culture and identity and a source of interest and of income.

NOTES

[1] Quotations are from local tourist brochures unless otherwise stated, notably the tourists' and visitors' comments.
[2] We are indebted to Ros Foskey for research assistance in Glen Innes.

REFERENCES

ANON. (n.d.) *Guide to the Australian Standing Stones*, Glen Innes and Severn Shire Tourist Association.
AUSTER, M. (1996) 'The Celtic imagination in exile', *Australian Folklore*, 11 (June), pp. 112–20.
BRENNAN-HORLEY, C., CONNELL, J. & GIBSON, C. (2007) 'The Parkes Elvis Revival Festival: economic development and contested place identities in rural Australia (with C.)', *Geographical Research* 45, pp. 71–84.
GIBSON, C. & CONNELL, J. (2005) *Music and tourism: on the road again*, Channel View, Clevedon.
GOTTDIENER, M. (2001) *The theming of America: American dreams, media fantasies and themed environments* (2nd edition), Westview Press, Boulder.
HARVEY, D. (1989) *The condition of postmodernity*, Blackwell, Oxford.
HOBSBAWM, E. (1983) 'Introduction: inventing traditions', in Hobsbawm, E. & Ranger, T. (eds) *The invention of tradition*, Cambridge University Press, Cambridge, pp. 1–14.
LOWENTHAL, D. (1985) *The past is a foreign country*, Cambridge University Press, Cambridge.
MAIR, H. (2009) 'Searching for a new enterprise: themed tourism and the re-making of one small Canadian community', *Tourism Geographies* 11, pp. 462–83.
MATHEW, J. (n.d.) *Poems of the stones*, Glen Innes.
PARADIS, T. (2002) 'The political economy of theme development in small urban places: the case of Roswell, New Mexico', *Tourism Geographies* 4, pp. 22–43.
PARADIS, T. (2004) 'Theming, tourism and fantasy city', in Lew, A., Hall, C. & Williams, A. (eds) *A companion to tourism*, Blackwell, Oxford, pp. 195–209.
PATTERSON, A. & BROWN, S. (2007) 'Inventing the pubs of Ireland: the importance of being postcolonial', *Journal of Strategic Marketing* 15, pp. 41–51.
RUTING, B. & LI, J. (2010) 'Tartans, kilts and bagpipes: cultural identity and community creation at the Bundanoon Scottish Festival', in Gibson, C., Connell, J. & Darian-Smith, K. (eds) *Festival places: revitalising rural Australia*, Channel View, Clevedon (in press).

Australia's Capital of Jazz? The (re)creation of place, music and community at the Wangaratta Jazz Festival

REBECCA ANNE CURTIS, *Candidate for Master of Studies in International Human Rights Law, University of Oxford, Oxford, United Kingdom*

ABSTRACT *An otherwise little-known country town has become a 'capital' for one creative scene—jazz. The Wangaratta Festival of Jazz plays a significant role in both nourishing the local community and Australian jazz music. The festival creates a unique space for performance, listening and interaction which intimately connects Wangaratta with major cities in Australia and overseas. Despite the fact jazz has no roots in Wangaratta, the town is increasingly significant in jazz circles as the pinnacle of musical excellence and integrity for 4 days of the year. The festival's impact lives on beyond the physical boundaries of the festival through new social connections, recordings and ensembles which are born out of it. In a very real sense, Wangaratta is Australia's capital of jazz because it has built a reputation as the place where jazz belongs.*

> You get a fantastic beautiful confusion of inner city bohos and farmers and freaks and drunkards and yobbos and intellectuals and international super stars ... it is the master stroke that stirs the fires of creativity and is just downright more interesting. Cities are boring on their own, as are rural places. But mix 'em up and you get the Wangaratta Festival which is testimony to the crucible appeal. (McAll, pers. comm. 2007)

Introduction

The role that music festivals have played in shaping cities and regions, and more immediately lives, is partly due to the ability of music to tap into our hearts and minds. It is also due to the power of music in building a sense of community or belonging within a particular musical style. The growth of popular music festivals around the world since the late 1960s (Woodstock, Monterey and the Isle of Wight Festivals) has produced not only music but also place, space and identities (Duffy 2000). In the genre of jazz, music festivals have created tourism sub-cultures at international jazz festival sites around the world (notably the Montreux, Montreal, North Sea and Newport Jazz Festivals).

The context within which music festivals occur is revealing. Music festivals define, represent and transform not only physical and material place and space but also the social, cultural and economic relations of people on many levels. The non-musical elements of festivals create spaces for social interaction, entertainment and enjoyment; built environments for the production and experience of music, such as halls and live music venues; cultural identities; and locations for various political and commercial intents (Gibson & Connell 2003). The musical elements of festivals—make-up of band, musicians, performances, tunes and approach—are equally important.

This paper focuses on Australian jazz music and examines the Wangaratta Festival of Jazz. The festival, held in the country town of Wangaratta in the state of Victoria, Australia, is Australia's premier jazz event. The paper draws on research conducted in 2007 and 2009 and offers insights from eminent Australian and international musicians, festival participants and others. It uses a range of research methods including: formal and informal interviews with musicians, festival organisers and local residents in Wangaratta; observations during visits to the festival in 2007 and 2009; and analysis of written material from books, journals and the media. As the festival reached its 20th year in 2009, it is timely to consider how this country town has been and continues to be a 'capital' for jazz in Australia. I suggest that what matters is the space the festival provides for the pursuit of musical excellence and social interactions which shape musical passions and pursuits—engendering different senses of belonging.

Making sense of jazz, festivals and place

Popular music festivals are known and valued for their ability to attract tourists, economic growth and development, and express social and political statements. They are, in the broadest sense, the oldest form of music tourism (Gibson & Connell 2005). Festivals provide places of spectacle and unique experiences, with one-off performances, networks of performers and audiences connected to particular musical genres (Connell & Gibson 2003; Hinton 1995; Gibson & Stewart 2009). However, there have been relatively few studies of jazz festivals (cf. Saleh & Ryan 1993; Formica & Uysal 1996), and none in Australasia.

Jazz first reached Australia in the 1920s and became popular as dance music, especially in the big band swing era. After the Second World War the jazz movement in Australia began to flourish. In December 1946, the inaugural Australian Jazz Convention was held in Melbourne, Victoria, which became a significant place for the development of jazz in Australia. In the early days of jazz in Australia, the music was closely aligned with the broader artistic scene. At the time, jazz was perceived as a radical, unconventional form of music and was an adjunct to the work of artists, painters and poets—the radicals of the time. Early jazz festivals in New South Wales, including the Deniliquin Jazz Festival, first held in 1970, and the Sydney-based Manly Jazz Festival which began in 1977, still draw crowds to the present day.

Today, jazz in Australia exists as a distinct scene. There is a focus on the individual pursuits of musicians and few are interested in cross-fertilising with other creative arts. This cultural shift has meant that jazz-oriented venues and scenes have become more vital to the survival of the music. Pianist Mike Nock suggests that jazz today remains a form of art, but its depth and substance are lacking:

> The talent and ability of young musicians coming up is quite astounding. But often, they don't understand the bigger picture. They are not even interested in the bigger picture and what ends up happening is that there is not a lot of substance at the end of it. Everyone kind of loses out. Because this is art. Whether it is folk art or high art it is all those things. In the end it depends on the people. But we are dealing with something a little deeper. For example, the best blues music deals with the very depth of things. It is all about depth. And that's the thing that I find lacking from so much of current trends in Australian jazz. (Nock, pers. comm. 2007)

Within Australian society, for many jazz is associated with the 'elite' or 'chardonnay set' (that is, the upper middle to upper class) in Australia's cities. For example, in 1996, 51 per cent of attendees at the Wangaratta Festival of Jazz were professionals (Wangaratta Festival of Jazz 2007). Given the income-earning power of this group, it is no surprise that jazz has been a vehicle for economic invigoration in rural towns in Australia (Gibson 2007). My argument here, however, is not so much about the earning potential of a jazz festival, but instead the role of jazz festivals in the creation and transformation of music and place. By creating a space for experimentation and valuing the social networks that underpin jazz, Wangaratta has fostered a sense of belonging, cementing its literal and figurative place in the Australian jazz scene.

Understanding the place of jazz festivals

While jazz remains primarily an urban phenomenon, the majority of jazz festivals in Australia are held outside major cities (Gibson 2007; Table 1). Cities support scenes and creative industries which form a critical mass of musicians, venues and structures that support the exchange of musical ideas, influences and sounds. Scenes are the infrastructure or mechanics that enable the creation and production of music to take place. Musicians and audiences everywhere reproduce ideas about the uniqueness of local music scenes which are born out of, and thrive in, urban places (Connell & Gibson 2003). Cities have become famous for their local approaches to jazz—from New York to New Orleans to London and Sydney. If urban scenes are a vital part of jazz music, how then, can we make sense of rural jazz festivals?

The objective of jazz festivals in rural and regional Australia has not been to create scenes or creative industries, but rather to establish social and cultural spaces for music to be used as a tourist product. Key to the establishment of these festivals has been the ability of places to market their location as an appealing physical, cultural and social landscape. Rural towns have used their mountains and forests, beaches and cliffs and vineyards and cheeses to create a sense of distinctiveness and attract people to a particular location to experience jazz and other localised cultural activities. Jazz festivals are increasingly woven into place promotion, for example, through the explicit linking of jazz sounds to place imagery, such as jazz 'in the vines' or 'in the pines' or 'on the rock' or in the 'valley' or 'at the farm', a practice that has contributed to the enhanced perception of such places as elite spaces. Whether or not a local scene exists, places have captured the 'sound' of jazz as an

TABLE 1. Australian jazz festivals by city or rural location

Australian jazz festivals	City	Rural or island location
2007 Valley Jazz Festival	×	
21st WA Jazz Festival		×
Articulating Space Festival of Exploratory Music	×	
Bellingen Jazz Festival		×
Big River Jazz Festival		×
Bronte Village Jazz Festival		×
Cool Nights, Hot Jazz	×	
Earthbeat Mildura Jazz		×
East End Jazz Festival	×	
Great Tropical Jazz Party		×
Half Bent Music Festival	×	
Hot Nights, Cool Jazz	×	
Illawarra Jazz Club—Easter Jazz Festival		×
Jazz: Now	×	
Jazz at the Farm		×
Jazz in the Vines		×
Kyneton Trad Jazz Festival		×
Manly Jazz Festival	×	
Melbourne Jazz Fringe Festival	×	
Melbourne Women's International Jazz Festival	×	
Noosa Jazz Festival		×
Norfolk Island Jazz in the Pines		×
Perth International Arts Festival	×	
Rhythm on the Rock		×
Shoalhaven Jazz Festival		×
Stonnington Jazz 08	×	
Sydney Festival	×	
TAC Wangaratta Festival of Jazz		×
Thredbo Jazz Festival		×
Wagga Wagga Jazz Festival		×
York Jazz Festival		×

Source: Jazz Australia (2009).

economic, cultural and social resource. Jazz has become a means by which places can be 'put on the map', transform themselves, and enhance their identity.

Wangaratta: (re)creating place, music and community

It's amazing when you go to this town, suddenly it becomes jazz land for four days of the year. (Hibbard, pers. comm. 2009)

The Wangaratta Festival of Jazz is located in a small town situated in north-east Victoria (see Figure 1) and is centrally located to Melbourne (2½ hours' drive), Sydney (7 hours' drive) and Canberra (4½ hours' drive). The festival has been acknowledged as a Tourism Victoria 'Hallmark Event' and Tourism Australia's 'most significant regional festival' (Rural City of Wangaratta 2009). The Rural City of Wangaratta, which encompasses the Alpine Valleys of the King and lower Ovens Rivers, has a total population of approximately 28 000. The town is one of Victoria's

PLATE 1. Sign at the entrance of the Rural City of Wangaratta.

growing regional provincial centres, through its industrial, retail and tourism sectors (Rural City of Wangaratta 2009). The urban centre of Wangaratta comprises 18 000 residents (Rural City of Wangaratta 2009). The region is known for its mountainous geography, gourmet food and is home to 35 of Australia's top wineries.

The festival was established in 1989 by a group of volunteer business people from the local community who sought to establish a landmark cultural event which would attract tourists to the town and boost the local economy (Wangaratta Festival of Jazz 2007). The event was not based on any pre-existing local jazz history or interest in jazz music, but was born from a feasibility study. The primary motivation of the organisers was to establish a credible event that provided the youth of the city

PLATE 2. Wangaratta town centre during the Wangaratta Festival of Jazz in 2009.

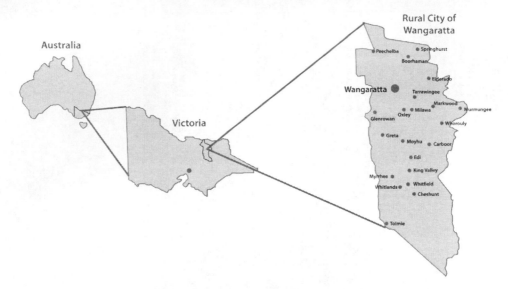

FIGURE 1. Map of Rural City of Wangaratta.
Source: Rural City of Wangaratta (2009).

with musical and tourism-related experience, while also generating revenue for the local economy through increased tourism (Clare 1999).

The festival is supported by various regional, State and Commonwealth government bodies and private businesses and organisations. Its major partners are the Transport Accident Commission (TAC), Arts Victoria, Tourism Victoria, the Rural City of Wangaratta, the Australia Council and the Australian Broadcasting Commission (ABC) (Wangaratta Festival of Jazz 2009b). The most visible of these partners has been TAC, from the branding of festival umbrellas on Reid Street to its advertisement in the festival program of a jazz musician playing a breathaliser like a trumpet. In what clearly was a creative move by festival organisers to keep festival goers safe, TAC has provided free breath testing and a public awareness campaign for festival attendees since 1993 (Wangaratta Festival of Jazz 2009a). The festival is also sponsored by Brown Brothers, Friends of the Festival (local businesses), the City of Stonnington, the Australasian Performing Right Association (APRA) and the Australasian Mechanical Copyright Owners Society (AMCOS) (Wangaratta Festival of Jazz 2009b).

Since its initiation in 1989, the festival has been run primarily by volunteers: an honorary management board and hundreds of local residents who mobilise during the event. It commenced in 1990 as a 2-day event and has grown to be a 4-day event of jazz and blues, showcasing more than 350 young and established Australian and international artists. Bob Dewar, who was closely involved in planning the festival, described its establishment as follows:

> We just thought that Wangaratta, the whole area, needed a kick, an infusion of money, some event ... you know of course that North East Victoria is a food bowl. Quality wines, meats, vegetables. Good food, good wines, good music. It all seemed to go together. When we decided on a jazz festival my idea was to appeal to the musicians, not the public! That may sound funny, but I felt that if it received credibility from the

PLATE 3. Performance by Sylent Running at the Performing Arts Centre, featuring winner of the National Jazz Awards at Wangaratta in 1990. Pianist-composer Barney McAll, Melbourne vocalist Gian Slater and United States guitarist Nir Felder, along with Chris Hale (bass guitar), Dan West (laptop) and Ben Vanderwal (drums).

> musicians the people would follow. They followed the musicians. That was how it would gain a reputation ... we would have a festival of credibility. (Dewar cited in Clare 1999, p. 38)

From modest beginnings, the Wangaratta Festival of Jazz has become the largest cultural event in north-eastern Victoria and is recognised nationally as Australia's premier jazz event.

The festival was created at a time when contemporary Australian jazz music was gathering momentum. Artists like pianists Barney McAll and Sean Wayland, guitarist James Muller, saxophonist Sandy Evans and trombonist James Greening were approaching jazz from a diverse and fresh perspective. Australian artists who have performed over the years at the festival include Don Burrows, James Morrison, alto saxophonist Bernie McGann, pianists Mike Nock and Paul Grabowsky and young artists such as singer Michelle Nicolle, saxophonist Matt Keegan, bassist Phil Stack and trumpeter Scott Tinkler. A range of established international artists and young overseas artists have also played at the festival.

The festival is held in several venues within the town, including a performing arts centre, a Blues marquee, local hotels and wineries, and a free stage on Reid Street. The majority of visitors to the festival in 2006 were from Melbourne (35 per cent) and elsewhere in Victoria (20 per cent) (see Figure 2). In the same year, over 30 000 individuals attended the 120 performance events at the festival (Wangaratta Festival of Jazz 2007). Growth in revenue from ticket sales from 1990 to 2007 has been impressive. In 1990, sales were approximately $25 000. By 1994 they had reached $80 000 and more than doubled to $190 000 by 1998. In 2006 the festival achieved its highest ticket sales in the history of the event, generating nearly $280 000. Revenue in 2006, including government grants, corporate and private sponsorship, and commissions from venues in sales, totalled $644 000.

The festival has also tapped into the region's rich environmental attributes, including wine and gourmet food. In 2006 the Milawa Gourmet Region, part of the Rural City of Wangaratta, used the area's reputation as a cultural centre to attract

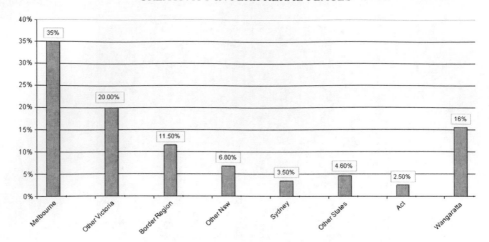

FIGURE 2. Place of residence by region of attendees at the
Wangaratta Festival of Jazz in 2006.
Source: Wangaratta Festival of Jazz (2007).

tourists to their very own 'Beat the Winter Blues 'n' Jazz Weekend'. The event took place in local wineries, cafés, restaurants and country pubs and its success can be attributed in large part to return visitors from the Wangaratta festival (Jackson, pers. comm. 2007). What matters most is the manner in which Wangaratta has fostered a sense of place and belonging through its jazz festival.

The significance of Wangaratta as location and social and cultural landscape

To understand how the festival may shape, and be shaped by, place and space, I explore its context—local musical practices, social institutions and cultural behaviour. Music plays a significant role in facilitating notions of community and collective identity, grounded in physically demarcated urban and rural spaces (Whiteley *et al.* 2004; Connell & Gibson 2005). According to pianist Mike Nock, music is only half of jazz, the other half being the social and cultural context in which jazz takes place (pers. comm. 2007). Wangaratta brings these elements together.

The festival plays a significant role in shaping local community relations across cultural, economic and social spheres. The launch of a new multimillion-dollar performing arts centre in October 2009 as the festival's primary venue, replacing the old Town Hall that served as the previous festival hub, is symbolic of how the festival has progressed since its inception.

The establishment of this new venue signifies the emergence of a new performing arts consciousness in the community and offers a gateway to a heightened sense of belonging and community in non-festival periods. Festival participants are not only an audience for the music but also of the local environment (people, venues, cafés, wineries and bars).

The festival is said to uplift the community, with the whole town getting involved in recognition of the festival's role in bringing life and business to the town (Clare 1999; Hibbard, pers. comm. 2009). Questions remain, however, about the extent to which the festival has a lasting local or regional impact in relation to jazz music. As co-Artistic Director of the Sydney-based Jazzgroove Association John Hibbard

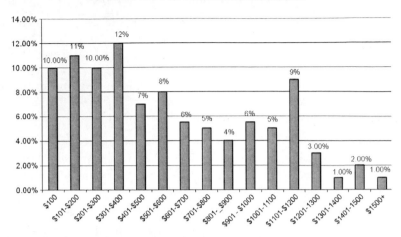

FIGURE 3. Estimated expenditure of attendees at the Wangaratta Festival of Jazz in 2006 (excluding festival passes and tickets) on accommodation, food, fuel, shopping, souvenirs and local products and produce.
Source: Wangaratta Festival of Jazz (2007).

explains: 'I expect it is very much like a circus, it arrives and then leaves and no one takes up juggling' (pers. comm. 2009). At a local level, the festival has encouraged more music programs in local schools but these programs have not focused explicitly on jazz music. Arguably, the festival has had a larger non-musical local impact through educating primary school students on road safety through its TAC sponsorship. Festival organisers have in the past attempted to shape local music culture through regular jazz performances in local pubs, without success. More-over, a TAFE program was established in a nearby regional town as a centre for music excellence in jazz, but was forced to fold owing to lack of local interest. As Festival Artistic Director Adrian Jackson asserts: 'In the beginning the locals were passionate about the festival. They were not in it for the music; they were in it for Wangaratta' (pers. comm. 2007). At a broader regional level, there is a performing arts presence through a strong concert band music camp tradition north of Wangaratta and music education and dance in Corowa (Hibbard, pers. comm. 2009).

The Wangaratta Festival of Jazz has had a significant impact on the local economy in the Rural City of Wangaratta. In 2006, 50 per cent of festival attendees spent more than $500 at the festival (excluding festival passes and tickets) on accommodation, food, fuel, shopping, souvenirs and local products and produce (see Figure 3). The Wangaratta festival, however, has not been central to the creation of a local cultural economy centred on jazz (in the sense of economic activity supporting the production and distribution of music). This is in stark contrast to Byron Bay in the Far North Coast, New South Wales, Australia, where its identity as a musical mecca and 'alternative' lifestyle location was born out of the annual 'Blues & Roots Music Festival':

> Music has been central to the creation of a unique cultural economy in Byron Bay. This has been bound up in a demographic shift—as migrants move to the region for lifestyle and amenity, constituting a new market for cultural products. (Gibson & Connell 2003, p. 182)

The emergence of an organic music movement, accompanied by a cultural industry, has become fully integrated into Byron Bay's character and local landscape. It is an international site where music is played, produced and lived. By contrast, in Wangaratta, jazz music is otherwise largely absent from the place of Wangaratta. But this does not mean the festival and its location are any less significant. Co-Artistic Director of the Sydney-based Jazzgroove Association Matt Ottingnon believes the festival's success is inextricably linked to the place of Wangaratta:

> The success of the Wangaratta Festival of Jazz is due to the fact it is in Wangaratta. Because it is a national jazz festival. If it was in Sydney or Melbourne, its successes would not be the same. (Pers. comm. 2009)

In other words, Wangaratta is unique and distinctly identified with the festival. Because Wangaratta is a small place the festival both absorbs many more townsfolk than at other times of the year, and has placed the town on the map.

Music matters

> It is a very special festival. It is the only festival in Australia that I know of that has such high music credibility. Most of the festivals are not run with this kind of integrity. (Nock, pers. comm. 2007)

> The great thing about it is that the organisers have given me [the Artistic Director] full creative power, within the allocated budget, to do whatever I like musically. They do the local community part, which I have nothing to do with, and I do the music. The two are mutually supportive of each other. (Jackson, pers. comm. 2007)

Does the music matter in Wangaratta? While the music's local impact is uncertain, what is clear is that for city-based jazz musicians, the festival has a sacred quality as a type of 'homeland' owing to the connective properties of jazz music. It bonds musicians in a similar way to diasporic populations, and the meaning and significance they share through an imagined spiritual homeland. Notwithstanding the obvious lack of a local scene, it is considered within jazz circles to as a special place not only for the creation of music but also as a kind of tribal meeting point where musicians can talk together, listen together, perform together, and drink together (Hibbard, pers. comm. 2009; Ottingnon, pers. comm. 2009). It is 4 days of the year when musicians gain an injection of inspiration and form enduring social and musical relationships. As Matt Ottingnon points out: 'it is closer than New York' (pers. comm. 2009). For pianist Barney McAll, Wangaratta is a place where musicians can catch up:

> Musicians are a strange lot and tend to isolate themselves in the sense that you might make a significant recording with some musos and then, never see them again, or you might go on the road all over the world and then when the band finishes, that is it. The 'family' is broken. However, at Wangaratta you can have a true catch up. (Pers. comm. 2007)

In this sense, through its commitment to artistic freedom, Wangaratta has enabled those within the jazz community to feel that they belong at this place and at this event, catalysing non-musical relationships and a sense of belonging in a way that further cements the event's reputation.

The festival's identity as a celebration of modern and contemporary jazz, combined with its reputation for musical excellence, has meant that it has acquired a competitive edge over other Australian jazz festivals and most credibility among artists. Musicians do not have to 'sell out' or 'water down' their music (Nock, pers. comm. 2007). The creation and implementation of the festival's musical identity by Artistic Director Adrian Jackson is widely considered critical to its ongoing success and credibility (Nock, pers. comm. 2007; Hibbard, pers. comm. 2009; Ottingnon, pers. comm. 2009; McAll, pers. comm. 2007). Beyond this the musical success of Wangaratta is underpinned by three key events. The first is the National Jazz Awards, an annual competition between young musicians in a given instrument. These awards distinguish Wangaratta from other jazz festivals by being recognised as the most important and prestigious jazz competition in Australia. The awards also provide a valuable opportunity for young musicians to meet, share ideas and receive exposure and recognition nationally. Second is the National Jazz Writing Competition, which showcases writing about or inspired by or responding to the music. Third is a series of youth workshops designed primarily for secondary-level students and other visitors in town for the festival. These events aimed at young musicians and writers form a critical structure through which the exchange of musical ideas takes place.

Additionally, the involvement of mainstream media, such as Australian Broadcasting Commission (ABC) radio, plays an important role in the credibility and attractiveness of the festival. The festival was not a natural success according to Artistic Director Adrian Jackson:

> Initially I didn't think it was going to work. I didn't think there would be a huge demand for modern and contemporary jazz in rural Victoria. In the first year we lost $50,000, although we received positive feedback. Some people believed in it and were willing to bank on its potential. (Jackson, pers. comm. 2007)

To the trained ear the music at the festival is considered 'authentic' and 'real'. According to Pianist Mike Nock at Wangaratta there is a pressure to play at your best:

> To me, I always have felt, ever since the beginning of Wangaratta, and I have been there since the beginning, that basically it is a place that demands that you give your best without any commercial considerations. That's how I have treated it. Not everyone would of course. It's a part of a contract you enter into with the audience. I feel totally that it is a space where musicians can be true to themselves and the music. (Nock, pers. comm. 2007)

Thus within the festival context, it is not in the musicians' interests to adapt their sound for tourists. They are 'true to the music' and are ambassadors of their own approach to jazz and advocates for the jazz movement as a whole.

Australia's capital of jazz?

> Wangaratta's location is a serendipitous master stroke on Adrian Jackson's part. Being between Melbourne and Sydney makes it perfect as a neutral place and the rural setting means you can actually play in a more open way. You are less laden with city life and complications and in a small way

I feel this is apparent in the music played at Wangaratta. (McAll, pers. comm. 2007)

One cannot overlook the fact that there is little jazz in Wangaratta outside the festival. But what Wangaratta does have is the power to capture musical credibility, authenticity and a sense of belonging for 4 days, in the one place, at one time, in a musical style that is facing threats to its survival in Australia and overseas. As John Hibbard explains:

> I noticed coming into Wangaratta that there are signs up saying it is Australia's jazz capital. It just can't be. Only for one week. But the amazing thing is there is probably more hard-core jazz going on there in that week than anywhere else, certainly regionally anywhere else in Australia, at any other time. (Pers. comm. 2009)

The musical experience at Wangaratta is insular and thoroughly urban. Jazz is played by musicians for musicians and on the whole by city people for city people in a type of musical vacuum which is not accessible to the average person who is not educated in music or jazz (Hibbard, pers. comm. 2009; Ottingnon, pers. comm. 2009). It is constructed as a festival of musical excellence, not a festival of easy listening. In other words, creating and filling a place with specific musical sounds (and the signs of credibility they convey) is crucial. The festival is a soundscape: a sonorous environment in which bodily, affective responses to sound are prioritised (Brant *et al.* 2007).

In musical experience, creative acts of perception engage the world, play a role in the social life of music and, to varying degrees, help to constitute society as a whole. Experience can be understood as the contents of consciousness: the ideas thought, the emotions felt, the sounds heard, the fragrances smelled, the textures touched and the colours seen (Berger 1999). Musical experience does not exist solely in performance or listening to sound but in the full range of settings (composing sessions, rehearsals, and listening events) where musical life is carried out and emotional relations established (Wood *et al.* 2007). The creation of musical phenomena is actively achieved through social practice, powerfully informed by the situation, the participant's goals in the event, and a potentially endless range of larger cultural contexts (Berger 1999).

To understand the musical experiences of listeners and performers at the Wangaratta Festival of Jazz is to make sense of its roots in jazz. Jazz music, by definition, is never performed live the same way twice. It is spontaneous or, in musical terms, improvised. Jazz contains new interpretations of the music based on the lived experience of the performer and how they experience the tune (emotionally, physically, on the stage). Improvisation can be understood as playing music from one's own perspective. In practice, it constitutes a new interpretation or approach to playing a tune loosely based on its original melodic or rhythmic form. Music is imagined and then communicated via improvisation—it is not read from a page (as in the case of classical music). As Pianist Mike Nock explains: 'I see myself as a painter. I don't start out with chords. I use colours and textures and let the music write itself, because it always tells you where it wants to go' (Nock, pers. comm. 2007).

Improvisation is the reason why people listen to jazz and why jazz is musically distinctive. It explains why musicians do not adapt their sound to meet tourist expectations and why listeners and performers engage with each other in a social context to share their perceptions of sound and its meaning. Improvisation also actively organises affective communication between the performer and the listener. Barney McAll describes that at Wangaratta this experience is intensified:

> Adrian Jackson has bridged the gap so that audiences can be happy with what they experience and the musicians are top notch and never compromise their sound. The wide variety, yet high quality, of musicians is key here. (McAll, pers. comm. 2007)

The event is thus experienced beyond the physical boundaries of the festival. The role of the festival space is matched by its role in the articulation of symbolic notions of belonging, community and musical experience which go beyond both place and time. The experience is 'here and now' and simultaneously within other places and times. As one example of one person's listening to one song on one night, the focus is on a particular listener's experience of the formal and emotional aspects of the song, the stage moves of the musicians, the taste of beer, and any other elements of the event that the participant concretely grasps. The participant's constitution of these experiences is influenced by his or her purposes in attending the event (the desire to get out of the house, to hear intricate and diverse music, to support the local festival) and his or her perceptual skills (listening skills, knowledge of music). These in turn are informed by past musical experience (years of listening to jazz, an interest in the Australian scene, music classes) and non-musical experiences (local environment, a good day), as Barney McAll elaborates:

> Wangaratta is a place where the psychological forecast can be told. Musicians travel around a lot and when they congregate and play they are describing their travels and experiences in abstract terms. That is why audiences often close their eyes at Jazz concerts. They are not as interested in the visual as they are in the story being told. The global update. The local issues in musical form. The bush poetry. (McAll, pers. comm. 2007)

In this way music then plays a significant part in the way that individuals author space, with musical texts being creatively combined with local experiences and sensibilities in ways that tell particular stories about the local, and impose collectively defined meanings and significance on space. Barney McAll is adamant that the music at Wangaratta is a description of life:

> The music at Wangaratta and the music I play is a description of life. Often musicians get caught up in rehearsing the potent music of the 60s in the United States but I feel many of the Australian groups are finding a new original voice. Adrian Jackson has picked up on this and presented it as jazz. The leather elbows and jazz stiffs can go and enjoy it too. (McAll, pers. comm. 2007)

> A lot of the greatest music festivals all over the world are in little out-of-the way places. You have to go somewhere to get there. You kind of get sequestered for the time in a different world. If you go from listening to one gig to another in the city it does not have the same intensity. Every year it is always great and the bar is constantly being raised. You will hear things that you will not hear anywhere else. (Nock, pers. comm. 2007)

The festival synthesises locally acquired musical knowledge from different scenes into a common place where music is articulated and shared amongst individuals within a collective identity. This ongoing process of music-making and listening is symbolic because at Wangaratta it achieves the status of a quasi or imagined scene in its own right. It is not a scene embedded in place but an imagined scene that transcends place on one level (a collection of scenes), but relies on it on another (the place of Wangaratta). When this is seen in the context of the struggle to find supportive venues for the expression of jazz in Australia's cities it becomes even more significant. The rich experiential settings in which the festival takes place overlap and intertwine to the extent that sounds produced by musicians in given local settings (i.e. their cities of origin) are deemed to have resulted from their sharing of music at Wangaratta. One of the festival's roles is to facilitate such diverse musical experiences. Australian jazz music is increasingly being played in a way that is more distinctly Australian and original. The festival has grown with more people and high-profile international groups participating in performances. Barney McAll describes this trend as follows:

> It is just deeply about music. It is the best festival to play at for this reason. There is no sense of compromise; of sucking up to corporations; it is just legit and I think people register and appreciate this fact. I would just say that Wangaratta becomes a barometer for me in a sense. I have played at the festival every two years since its inception and I get a vivid gauge of how my music has changed every time I play. I have also created more and more fans through playing Wangaratta and Adrian has been very supportive of all my whacked-out ensemble ideas. There is nothing that makes me prouder than playing the Town Hall when it's full of people who really enjoy my music and its feels like a net below a tightrope that I can never injure myself on. (McAll, pers. comm. 2007)

Conclusion

Jazz music is not indigenous to Wangaratta, yet Wangaratta is the lifeblood of Australian jazz music. The Wangaratta Festival of Jazz has become the means for an otherwise little-known country town to become Australia's jazz capital, and to develop into a mecca for jazz in Australia, whereby the music is re-routed from cities to the country for '4 days of peace and music'. The festival can be credited for creating and fostering musical and social relations between city and rural populations in Australia and internationally. It offers freedom for people to perform, talk, and 'do what jazz musicians do'. The festival's impact lives on beyond the physical boundaries of the festival through new social connections, recordings and ensembles which are born out of it.

The festival simultaneously enables community, stimulates creativity and encourages hybridity without compromise and without tensions between the musical and local communities. At a local level it plays a key role in community building and economic invigoration. At a broader level, it provides an event where performers and listeners can experience jazz in an intense and uninterrupted environment. The event allows for music-making across scenes (usually defined by their location within cities) and in turn leads to an imagined scene where ideas and concepts can be exchanged and experienced. Critically, it provides a place for the spontaneous and high-level production of music and acts as a source of inspiration to Australian jazz musicians.

As the Wangaratta Festival of Jazz shows, the role of festivals in connecting major cities with remote and smaller places is important and requires more credit from scholars of creative industries. It also challenges us to gain a better understanding of the interactions between tourism and jazz, and tourism and popular music festivals. Local music scenes in cities are constructed conceptually, socially and symbolically and, as in Wangaratta, involve linkages to festivals in remote and smaller places. This paper also highlights the need for scholars writing about creative industries to give more attention to music and how it is affectively experienced (Wood *et al.* 2007). Jazz music is a sonorous experience—it is about complex musical theory, imagination, spontaneity, and improvisation. Recent work blending insights from the social sciences, particularly human geography, with musicology, the perception and phenomenology of musical experience and insights from performers provides a model for such future research (e.g. Brant *et al.* 2007). Progress in this regard will be critical if we are to unpack how musicians work within particular spaces and places and how imaginative and emotional worlds shape and are shaped by specific musical practices, social interactions and sites.

REFERENCES

BRANT, R., DUFFY, M. & McKINNON, D. (eds) (2007) *Hearing places: sound place time and culture*, Cambridge Scholars Publishing, Newcastle.

BERGER, H. (1999) *Metal, rock, and jazz: perception and the phenomenology of musical experience*, University Press of New England, London.

CLARE, J. (1999) *Why Wangaratta?—the phenomenon of the Wangaratta Festival of Jazz*, Wangaratta Festival of Jazz, Rural City of Wangaratta.

CONNELL, J. & GIBSON, C. (2003) *Sound tracks: popular music, identity and place*, Routledge, Milton Park.

DUFFY, M. (2000) 'Lines of drift: festival participation and performing a sense of place', *Popular Music* 19, pp. 51–64.

FORMICA, S. & UYSAL, Y. (1996) 'A market segmentation of festival visitors: Umbria jazz festival in Italy', *Festival Management and Event Tourism* 3, pp. 175–82.

GIBSON, C. (2007) 'Music festivals: transformations in non-metropolitan places, and in creative work', *Media International Australia incorporating Culture and Policy* 123 (May), pp. 65–81.

GIBSON, C. & CONNELL, J. (2003) 'Bongo fury: tourism, music and cultural economy at Byron Bay, Australia', *Tijdschrift voor economische en sociale geografie* 93(2), pp. 164–87.

GIBSON, C. & CONNELL, J. (2005) *Music and tourism: on the road again*, Channel View Press, Clevedon.

GIBSON, C. & STEWART, A. (2009) *Reinventing rural places: the extent and impact of festivals in rural and regional Australia*, University of Wollongong, Wollongong.

HIBBARD, J. (2009) Interview.

HINTON, B. (1995) *Message to love: the Isle of Wight Festivals, 1968–1970*, Castele Communications, Chessington.

JACKSON, A. (2007) Interview.

JAZZ AUSTRALIA (2009) 'Festivals', available from: http://www.jazz.org.au/directory/festivals?regions[Region]=&festivals[select][FestivalID]=32 (accessed 14 October 2009).

McALL, B. (2007) Interview.

Nock, M. (2007) Interview.

Ottingnon, M. (2009) Interview.

Rural City of Wangaratta (2009) 'Discover—adventure, legends & indulgence', available from: http://www.wangaratta.vic.gov.au/CA256B5800826065/page/About+Our+Region?OpenDocument&1=101-About+Our+Region~&2=~&3=~ (accessed 15 October 2009).

Saleh, F. & Ryan, C. (1993) 'Jazz and knitwear: factors that attract tourists to festivals', *Tourism Management* 14, pp. 289–97.

Wangaratta Festival of Jazz (2007) *Research report: patron attendances*, Wangaratta Festival of Jazz, Wangaratta, April.

Wangaratta Festival of Jazz (2009a) 'Avoid being a bloody idiot', available from: http://www.wangarattajazz.com/cms-the-festival/tac-message.phps (accessed 5 December 2009).

Wangaratta Festival of Jazz (2009b) 'Supporters', available from: http://www.wangaratta-jazz.org.au/cms-the-festival/sponsors.phps (accessed 5 December 2009).

Whiteley, S., Bennett, A. & Hawkins, S. (2004) *Music, space and place: popular music and cultural identity*, Ashgate, Aldershot.

Wood, N., Duffy, M. & Smith, S.J. (2007) 'The art of doing (geographies of) music', *Environment and Planning D: Society and Space* 25(5), pp. 867–89.

Creative Migration: a Western Australian case study of creative artists

DAWN BENNETT, *Curtin University of Technology, Perth, Australia*

ABSTRACT   *It is well known that a pilgrimage overseas can be crucial to the career development of specialist creative artists. All too often, however, the pilgrimage becomes a permanent migration. Significantly, the loss of this creative talent is not limited to the national level. The dominance of cities as the centres of Australia's knowledge-based economy leads also to migration of creative artists from regional centres and from smaller cities such as Perth, lessening the potential for those regions to attract and retain creative and innovative people. Given the globalised nature of the cultural industries and the emergence of new technologies, this study of Western Australian creative artists considers whether migration loss could be repositioned as cultural gain. Initial results suggest that spatial separation due to geographic isolation is particularly problematic for Western Australian creative artists both within the regions and the metropolitan area. Despite participants' strong personal connections with Western Australia, artistic connections were tenuous and artistic involvement was negligible. Implications include the need to actively engage with creative migrants by fostering their continued involvement in the cultural life of cities and regions.*

Introduction

Creativity, whether that of artists or of those whose creativity is embedded within other areas of the economy, is often cited as crucially important to the economy (Anheier & Isar 2008; Glaeser 2003). However, many such claims derive from place-competition strategies such as Florida's theory of a creative class (2002), which somewhat simplistically positions creative artists as magnets for creating 'cool city' images (Peck 2005) to generate in-migration of creative and talented thinkers.

Since the creative industries discourse began with Keating's 1994 Creative Nation policy statement (Commonwealth of Australia 1994), the arts have received increasing attention within discussions of human capital, regeneration, community engagement, branding and image (Markusen & Gadwa 2009; Pratt 2009; Vanolo 2008). These alignments have fuelled a plethora of reports on the societal and economic roles played by the arts; however, little of this research documents the practice, migration and working lives of individual creative artists. The deficiency relates partly to the sheer variety of scale, economic activity and organisation of the creative industries, which Hartley (2005, p. 26) describes as 'an entirely new species of cultural and economic enterprise ... whose shape and extent has yet to be properly mapped and understood, even by the people involved'. The complexity of artists' work is such that it is not captured by national data collections that fail to take into account multiple employments (see Brennan-Horley, this issue). Indeed, Throsby suggests that Australian census data underestimate the artist population by over 50 per cent (Throsby 2008a; Throsby & Hollister 2003), and Pratt (1997, p. 1954) agrees that intensive analysis of the sector can be achieved only through empirical research, given that 'original data sources are weak and inadequate for the purpose'.

There is, similarly, little research on the migration of creative artists (Houston *et al.* 2008). The complexities of creative labour research are compounded in this respect because artist migration can be temporary, and temporary migration is generally an under-researched area yet to be sufficiently recognised 'by either researchers or demographic and labour data collection agencies' (Hugo 2006, p. 212).

Careers featuring multiple employments and limited opportunities within complex markets remain significant influences among the artist population, and a pilgrimage overseas remains a logical feature of career development. Many present-day creative artists will travel in order to develop their careers, reach larger markets, and work within established industries elsewhere in the world: 'the nomadism of artists is naturally a concerted action of moving and settling in order to discover and to create, in order to renew one's awareness and one's formal responses' (Haerdter 2005, p. 7). This migration is largely accepted by the creative sector, especially when the move is the result of a significant career opportunity (see Warren and Evitt, this issue).

Whilst artistic success is celebrated, however, the physical loss of creative artists and their work is often lamented by artists' places of origin, especially when the travel becomes permanent or long-term migration. Just as creative artists migrate overseas, the dominance of cities as the centres of Australia's knowledge- or experience-based economy leads to migration from regional centres and from smaller and more isolated cities such as Perth. If creativity is believed to be central to economic vibrancy, this out-migration lessens the potential for these areas to retain creative capital and sustain economic growth.

Given the globalised nature of the cultural industries and the emergence of new technologies, this study sought to understand some of the factors influencing artist migration and asked in what ways migration loss could be repositioned as cultural gain.

Approach

This paper reports results from a study of specialist creative artists practising within the 'core' creative industries (Throsby 2008b) of music, visual arts and film. Specifically, the study provided a snapshot of Western Australia's creative diaspora: creative artists living and working overseas or in the Eastern states of Australia. Participants, cited here using pseudonyms, included instrumental and vocal musicians, a conductor, a producer, a composer and a visual artist (five females and three males). Although this was a small sample, the depth and consistency of responses led to a wealth of information.

To avoid sample bias, participants were recruited from multiple independent sources including professional networks, alumni listings and industry press. Participants used e-mail to answer the research questions and clarify responses. Inductive coding was employed in the analysis, and an observer independently coded responses to ensure consistency and avoid bias. Respondents were invited to answer any or all of the following four questions:

(1) Do you still feel a connection with Western Australia? If so, in what ways?
(2) How did leaving WA impact your career?
(3) Are you likely to return to live or to visit in the future?
(4) What could be done to support WA's creative people and the cultural environment?

The study was informed by the results of research involving 143 classical instrumental musicians (Bennett 2008a), which utilised a survey to gather information on career trajectories, education, location, and working patterns. Relevant findings from this formative study, taken from the responses of musicians who had migrated during the course of their careers, are incorporated into the following summary and discussion. The discussion section also draws preliminary results from a current study of the career ambitions of music and dance students. The discussion is structured according to the three emergent themes of *cultural image*, *creative migration* and *artistic connectivity*.

Results and discussion

Cultural image

It is believed that the potential for a place to attract and retain creative and talented thinkers is impacted by the richness, vibrancy and diversity of its cultural environment. Not only is creativity linked to economic success, it is a 'fundamental means through which places are perceived' (Gibson *et al.* 2002, p. 174). Elise, a WA composer and singer now living in Berlin, noted 'incredible talent in Perth', which Melbourne-based visual artist Theo attributed to the 'unique upbringing that the unique setting of Perth allows'. However, positive comments about WA lifestyle that is 'the envy of most' (Theo) were counter-balanced by perceptions of a poor cultural image.

Despite Perth's consistently high scores in international liveability surveys, the image of Perth as 'Dullsville' is perpetuated within local media. This was taken up most avidly by Theo, who wrote:

> The cultural image of Perth to the rest of Australia is of a kitsch frontier town with little sophistication. This image is hard to transcend. The 'West Australian', 'Burkie', 'WA Inc.', 'Ben Cousins', 'Worst of Perth' etc. is a source of much amusement. Therefore this image needs to be addressed ... young visual artists feel the need to leave Western Australia as the cities of Melbourne, Sydney and increasingly Brisbane seem to be the centres of cultural activities and venues.

The creative migrants were asked whether they still felt a connection to WA. This was a deliberately broad question designed to elicit responses about both personal and professional connectedness. All of the participants expressed a strong personal connection with WA, typified by jazz performer Mia, now base in Sydney: 'Still feel like the laid back, sand-under-my feet, sun-on-my-shoulders West Australian gal. Wish I could live it and sustain my passion at the same time.'

Whilst the WA lifestyle was reported positively, a consistent theme was the different modes of practice required to sustain an arts practice in a smaller centre. Migrating back to WA despite feeling that there are 'more orchestras and performance opportunities in one corner of Hamburg than in the whole of Western Australia', one participant from the musician study (Gail) described the necessity to become self-employed: 'I chose to come back to Australia and I discovered within a very short time that I wasn't going to have enough performing to keep myself happy, so I knew I was going to have to create it. So I did.' The presence of an established music industry in WA made it possible for Gail to adapt her practice. For other artists, such as the TV producer discussed below, the lack of a local industry makes it impossible to return: she 'couldn't move to Perth, a city without a sitcom, without changing careers'.

The perceived vibrancy of a place, or what Jess and Massey (1995) term a 'mental sense of place', is influenced by the visibility of its activities. Despite Perth's reputation for a unique popular music scene (Stratton 2008), the city centre does not have the cultural heart or hub of many other cities; hence many cultural activities can be invisible to the uninformed (see Felton *et al.*, this issue). A study conducted for the Department of Culture and the Arts (DCA 2008) concluded that locating a list of WA and even Perth-based events was impossible without prior knowledge of where to look. The study recommended the formation of a website hosted by, or with a direct link from, Tourism WA (Bennett 2008b). A simultaneous survey conducted by the City of Perth found that image and vibrancy are also a concern for local residents. The survey attracted 600 responses and identified 15 major issues, three of which have particular relevance to the visibility of cultural activities:

- Development, attraction and support of creative and cultural activities;
- The need to enhance entertainment, dining and retail activities to increase the vibrancy and liveability of the city; [and]
- The need to focus on all aspects of the visitor experience to create a real 'buzz' for both interstate and overseas visitors. (City of Perth 2009, n.p.)

An initiative to arise directly from the City of Perth study is the proposed 'What's on in Perth' website. Although the website will initially profile only events within the city, it promises to be an important initial step towards bringing together Perth-based and regional events into a searchable database for residents and visitors.

Creative migration

Real-time or performative arts practice has distinct implications for mobility because of the physicality and transience of the practice. I drew on data from the musician research to ascertain the commonality of artist migration within a larger group of real-time creative artists. The 143 musicians were found to have studied in 208 locations including the USA, Europe, the Middle East and Australia. Of the 105 respondents based in WA, 38 per cent were migrants and a further 29 per cent had migrated for work and study. The music study found a similarly high proportion of international artists within orchestras: 30 per cent of orchestral participants had migrated from other countries and a further 27 per cent had lived or studied elsewhere. Results are consistent with Jang's (1996) findings that approximately one third of Korea's orchestra members come from outside of Korea.

For WA creative artists, who are closer to Southeast Asia than to Sydney and Melbourne, the difficulties of maintaining a sustainable creative practice is exacerbated by the wide spatial distribution of arts activity that deems intrastate, interstate and international touring prohibitively expensive (see Gibson *et al.*, this issue). There is almost twice the rate of cultural industries occupation in Perth as there is in the balance of the state and WA has the highest metropolitan primacy in Australia (Gibson *et al.* 2002). The vast distances between major centres presents a challenge for both independent and employed artists, such as orchestral musicians for whom changing orchestra necessitates moving state. None of the orchestral musicians in the musician study expected to remain with their orchestras, and many expected to leave behind family and friends in order to progress their careers. However, migration interstate or overseas becomes increasingly problematic as family and social ties are developed. Some of the musicians suggested that these barriers had become overwhelming: orchestral musician Deborah described the pressures of travel as being 'incompatible with growing family commitments', whilst drummer Matt found that 'the economic realities of sustaining a family did not agree with a life on the road or raising children in the comfort they deserve to experience'. Shifting priorities arose as an important factor:

> When once I was yearning for the illusive celebrity status so common to many teenagers, my ego seems to have finally receded to the point that I can now see who I truly am. I love performing, but the quest is no longer associated with fame. (WA-based pianist, Beth)

It is likely that similar factors exist across the arts, which raises the question of how much artist migration is reactive and how much pre-meditated. The extent to which future artists intend to travel or migrate is being investigated in a study of career hopes and expectations. The study has so far surveyed 186 first-year undergraduate music and dance students in WA, asking the students to consider their careers up to 5 years after graduation. Initial results present a startling picture. A random sample of 40 surveys reveals that 83 per cent intend to work in the Eastern States or overseas. Thirty-five per cent of the students plan a permanent migration.

Although this is merely an indication of career ambition, the results align with the level of migration reported within the music study and strongly suggest that migration is commonly planned prior to the commencement of professional practice. Drawing on a 'global imagination' (Rizvi 2000), a key driver for artists is the desire to make a mark within the imagined buzz of larger or exotic markets, increasing their status through national or international activity. Heightened status is illustrated in research conducted by Gibson (2003, p. 210), who noted that visiting artists can earn more because of their outside identity: 'mileage was made out of the credible or exotic nature of those [musical] products sourced from other locations'.

Whilst migration can be pre-meditated, the research also found much artist migration to be a reactive, reluctant move. The primary driver for artists was, in fact, a lack of local opportunities:

> I made some attempts to find employment with Western Australian film and television companies but quickly found that there were very few opportunities ... When I moved to Los Angeles I had hundreds more job options available to me. (Greta, TV producer living in California)

David, a professional trombonist, reflected:

> By moving to Melbourne I was exposed to a greater population of musicians, venues and audiences, therefore my performance opportunities increased. Being geographically closer to other cities also opens up more performance opportunities.

Carla, now a lead opera singer, gained continuous employment once she migrated to Germany:

> The year I auditioned at WA Opera there was only one position in the opera chorus going, and ten times that amount were applying for it—just to sing in the chorus [positions contracted on a seasonal basis].

Writing from Berlin, composer and jazz singer Elise made the connection between opportunities and venues: 'I'm sure I will visit Perth again but it is difficult to live there because of its limited venues and opportunities to play live.' Berlin is a particularly good example of a city that has embraced creative industry activities since the Cold War era, utilising derelict buildings to establish studio spaces and a vibrant cultural sector: 'Berlin's government and its people have taken a bet on creativity by plugging the gap in the economy as the city is rebuilt, and it is paying off' (Heath 2009, p. 140).

Dellbrügge and de Moll (2005) interviewed 30 artists in an attempt to identify a topology of attitudes and behavioural patterns among artists who had migrated to Berlin. They found that 'there is a necessity—even in Berlin—for most visual artists to earn money with odd jobs to finance their career' (p. 140). Further, 'artists may not necessarily benefit, as content-providers and generators of the creative setting, from the successful city branding through culture' (p. 142). Nonetheless, the Berlin artists cited pull factors including low cost of living, increased status due to their Berlin identity and, crucially, opportunities for critical debate among both artists and the general population. The presence of so many artists enabled them to feel that their work and working patterns were normal: 'frees them of the pressure to play the role of exotic creatures' (p. 141). Similarly, Drake, in his study of UK artists, found that 'locality-based intensive social and cultural activity may be a key source of inspiration' (2003, p. 522). Drake's respondents referred to 'the "buzz", unpredictability or excitement' (ibid.) of a specific city location in much the same way that the WA students imagined Eastern States and overseas locations, and the artist sample had imagined major centres such as Berlin and Melbourne.

Artists in many smaller centres require a 'fluidity of creative identities' (Luckman *et al.* 2009, p. 628) to sustain their arts practice. Even with multiple identities, the availability of local opportunities can be critical: as trombonist David explained, many WA artists 'have to work other jobs to fund the time for their passion; however in large cities many artists can work full time'. Melbourne-based pianist Jenny observed that 'the population of Perth and lack of proximity to fairly large regional towns and centres' can render full-time arts practice impossible and can confine artistic practice to what Theo described as 'a fringe activity'. This is exacerbated in WA by the high cost of travel, which often means that the cost of a regional tour far exceeds that of an interstate tour that could reach much larger markets. Viability is also threatened by the demise of venues, the result of multiple factors including the gentrification of inner city areas including key entertainment districts (Johnson & Homan 2003). With most artists working across multiple genres, sectors and in various situations (Markusen *et al.* 2006), venue demise has been felt across the visual and performing arts. Ironically, in this era of increasing artist independence, exposure via performances and exhibitions is vital to maintaining market share, being visible and remaining connected.

Artistic connectivity

Despite a desire among study participants to be involved in WA arts activities, artistic connections were tenuous and artistic involvement was negligible. Jazz performer Mia had attempted to remain involved: 'I used to try to organise gigs but it became too difficult for effort versus return.' Perth, she wrote, needs

> ... more public venues/community events that support non-mainstream style entertainment, government support for innovative projects. Promotion of West Australian artists to greater Australia and the world. I've always felt some sort of national 'gig circuit' or 'exchange' performance set up would stimulate performers and audience alike. Some sort of public creative 'hub' like you might find at somewhere like Federation Square in Melbourne ... where general public are regular exposed to music, art, acts they might not normally get to see/hear.

Jenny (pianist) had tried to organise Perth concerts from Melbourne in order to remain involved in the WA arts scene. However, she had found that:

> ... often it feels like that if one wants to stage anything, it has to be individually organised, managed and put on, without support ... The difficulty in finding appropriate and prestigious venues lessens the incentive for artists to showcase their talents, particularly as most of the prestigious venues with good facilities usually put on concerts under larger banners.

Similarly, Gavin, a big-band leader now living in Norway, wrote of Western Australia's 'extremely high level of musicianship', but he conceded: 'I won't be living there again. As much as I would like to, there are too many limitations.'

Two respondents believed that their migration had resulted in their being less than welcome within the 'local' arts scene. For example, Berlin-based singer Carla, wrote:

> I feel connected to the WA music scene, but I wish I was more connected ... I think they [name of company] have the policy too, of

supporting performers who have chosen not to leave Perth. I hope this will change in time.

Again reflecting the need for inter- and intraconnectedness, jazz singer Elise wrote of the 'open, enthusiastic attitude' she had encountered in other cities and suggested: 'community, students, teachers, and people who just want to be involved need to stick together, especially in a small city like Perth'.

The WA cultural industries are characterised by sparse, sometimes problematic communications between multiple networks. This fragmented communication is typical of a sector founded on grassroots community-based social networks. The desire to connect artistically emerged as a significant theme among the creative diaspora and suggests that virtual artistic collaborations and networks would have artists' support and engagement. Whilst local critical mass is not something that can be developed quickly, the creative migrants highlighted the importance of connecting with national or global industry networks, and the difficulties they had faced in this regard when practising in WA:

> ... the performing and creative arts industries are ones that benefit largely from 'who-you-know' syndrome ... furthermore, large touring acts do not always come to the West coast if they tour Australia. Having exposure to international artists in the Eastern States broadens and deepens one's perspective. (Jenny)

Most of the respondents alluded to the 'invisibility' and professional isolation they had experienced when they had based their practice in WA: 'it is very difficult to transcend the regionalism and participate in the national conversation about visual culture' (Theo). This comment reflects the importance of critical debate, which was such a draw card for the Berlin artists. Isolation was in part aligned with the need to be aware of opportunities as they arise; and yet, as Greta wrote from California, the solutions can be quite simple: 'through my UCLA [University of California, Los Angeles] writing course ... I received a weekly email with a hundred or so entry level jobs'.

Several participants noted that smaller centres require a greater entrepreneurial approach to creating and managing performance or exhibition opportunities. The entrepreneurial skills required to build sustainable arts careers are often absent or underdeveloped in the formative stages of artists' careers. This impacts the ability of emerging artists to creative a viable practice, particularly in smaller centres where more opportunities need to be self-generated. The impact of migration on creative artists who migrate early in their professional careers, when friends and family can be critically important, is an issue deserving of further research. Theo made the point that a smaller market can provide valuable opportunities for early career development. He reminisced:

> I left as a young adult, and I was surprised at how moving to another city as a young artist was with difficulty. I had to form new social and support networks and this did take years to form. Ironically, I feel that I missed out on opportunities that colleagues in Perth were allowed, in particular grants and commissions. Therefore my early development through opportunities as an artist may have been stunted.

One of the obvious strategies for retaining and engaging emerging artists is to connect them with their professional fields whilst they are still training. Although this occurs frequently within some of the visual arts and within contemporary music, in other fields it is common for new graduates to have few or no professional connections, as was the experience of TV producer Greta:

> When I graduated I wasn't given any instruction on how to use my film studies in Western Australia. In three years of university I hadn't met anyone who was actually working in production or post-production in Perth. I didn't study Western Australian films, attend any local film festivals or learn anything about making films or television in small creative communities like WA. Essentially I knew nothing about the workings of the Perth entertainment industry and, as a result, quickly decided to move to away from the State.

Pianist Jenny noted the initiative of Melbourne radio station 3 MBS, which supports a resident artist program and weekly lunchtime classical music concerts at which young performers and performance students perform live-to-air from the studio, providing 'invaluable experience, something to work for and, as far as I can see, is mutually beneficial to all involved'. Relatively simple initiatives, such as shared bulletins, live-to-air concerts and exhibition or performance spaces within city buildings and public spaces, produce both opportunities and vibrancy. They also contribute much-needed venues.

Concluding comments

The migration of creative artists is driven by push factors (Lee 1966) such as limited local opportunities and geographic isolation, and pull factors including experience, identity, networks, and the draw of more established industries or geographic clusters of activity.

Despite some alignment, it is clear that research on artist migration, including the spatial distribution of artists within metropolitan areas, is not generalisable from the results of research on the creative class. Nor do existing primary data provide a reliable base for such research. Recent research from Sweden (Hansen & Niedomysl 2009) suggests that migration among the creative class (used as a proxy for human capital) is driven primarily by employment, closely followed by friends and family. Creative artists similarly migrate for employment; however, the move is rarely the result of securing a position. Rather, it tends to be a move to a location where there are increased opportunities to create work, either because of the presence of an established industry, or simply because of a larger population base. This is often an unstable migration involving financial risk. Another important difference is that whereas the creative class is drawn towards friends and family, artists tend to move away from important social networks in order to further their artistic practice. In a sector where increased success often equates to increased time away from home, artists moving towards friends and family often do so at the expense of their practice.

Given that artists can be creatively generative in terms of the way in which a place is imagined, their involvement in both the process and project of globalisation is a largely unrealised asset little celebrated within the local context. Using the examples of the Australian-derived *Village Roadshow* and *Hopscotch Films*, O'Neil

(2009) observes that the 'commercial success of Australian companies in the creative economy is scarcely acknowledged here' (p. 32); and yet the success of such creative endeavours could form an important component of how their place of origin is imagined.

Perhaps this is an element of migration and identity that has shifted in recent times. In 1983, Anderson suggested that communities 'are to be distinguished not by their falsity/genuineness, but by the style in which they are imagined' (p. 6). However, this idea has less credence in the context of societies imagined from a global perspective: in fact, the project and process of globalisation may now reinforce that which is locally believed and thought about (Moore and Held 2007). Whilst globalisation and advanced technologies can be imagined as threats to local cultures and community identities, they offer new opportunities to retain long-term, tangible links between creative artists and their places of origin. Virtual networks are invaluable in fostering much-needed and valued critical debate, and much could be done to connect local artists with national and international networks including those involving creative migrants. This would enhance the value of experience brought back to the local context.

The study provides a snapshot of the drivers influencing artist migration. It suggests the value of actively engaging with creative migrants to foster their continued involvement as active agents in the cultural life and image of our cities and regions. A targeted program of virtual collaborations could go a long way towards connecting artists and community regardless of artists' locations. This would reposition creative migration as a more positive element of local identities.

REFERENCES

ANDERSON, B. (1983) *Imagined communities: reflections on the origins and spread of nationalism*, Verso, London.

ANHEIER, H. & ISAR, R. (eds) (2008) *Cultures and globalization: the cultural economy*, Sage, London.

BENNETT, D. (2008a) *Understanding the classical music profession: the past, the present and strategies for the future*, Ashgate, Aldershot.

BENNETT, D. (2008b) *Somewhere to play: venues and live original music in Western Australia*, Department of Culture and the Arts, Perth.

CITY OF PERTH (2009) *City of Perth 2029: we hear you!*, City of Perth, Perth.

COMMONWEALTH OF AUSTRALIA (1994) *Creative nation: Commonwealth cultural policy*, Department of Communications and the Arts, Canberra.

DELLBRÜGGE, C. & DE MOLL, R. (2005) *Artist migration Berlin*, Heidelberger Kunstverein, Berlin.

DEPARTMENT OF CULTURE AND THE ARTS (2008) *Arts and culture in Western Australia*, DCA, Perth.

DRAKE, G. (2003) ' "This place gives me space": place and creativity in the creative industries', *Geoforum* 34, pp. 511–24.

FLORIDA, R. (2002) *The rise of the creative class*, Basic Books, New York.

GIBSON, C. (2003) 'Cultures at work: why "culture" matters in research on the "cultural" industries', *Social and Cultural Geography* 4(2), pp. 201–15.

GIBSON, C., MURPHY, P. & FREESTONE, R. (2002) 'Employment and socio-spatial relations in Australia's cultural economy', *Australian Geographer* 33(2), pp. 173–89.

GLAESER, E. (2003) 'The new economics of urban and regional growth', in Clark, G., Feldman, M. & Gertler, M. (eds) *The Oxford handbook of economic geography*, Oxford University Press, Oxford, pp. 83–98.

HAERDTER, M. (2005) 'NeMe: remarks on modernity, mobility, nomadism and the arts', presentation given at Res Artis, Cyprus, April, available from: http://neme.org/main/137/nomadism (accessed 25 May 2009).

HANSEN, H. & NIEDOMYSL, T. (2009) 'Migration of the creative class: evidence from Sweden', *Journal of Economic Geography* 9(2009), pp. 191–206.

HARTLEY, J. (2005) 'Creative industries', in Hartley, J. (ed.) *Creative industries*, Blackwell, Malden, MA, pp. 1–40.

HEATH, R. (2009) 'Risky business in challenging times', in Schultz, J. (ed.) *Essentially creative*, *Griffith Review Autumn 2009*, ABC Books, Sydney, pp. 137–46.

HOUSTON, D., FINDLAY, A., HARRISON, R. & MASON, C. (2008) 'Will attracting a "creative class" boost economic growth in old industrial regions? A case study of Scotland', *Geografiska Annaler B* 90, pp. 133–49.

HUGO, G. (2006) 'Temporary migration and the labour market in Australia', *Australian Geographer* 37(2), pp. 211–31.

JANG, K. (1996) 'Dear friends! Let us start over', in Oliva, G.M. (ed.) *The ISME Commission for the Education of the Professional Musician 1996 seminar. The musician's role: new challenges*, Universitetstryckeriet, Lund, pp. 25–136.

JESS, P. & MASSEY, D. (1995) 'The conceptualization of space', in Massey, D. & Jess, P. (eds) *A place in the world: places, cultures and globalization*, Oxford University Press, Oxford, pp. 45–85.

JOHNSON, B. & HOMAN, S. (2003) *Vanishing acts: an inquiry into the state of live popular music opportunities in New South Wales*, Australia Council and the NSW Ministry for the Arts, Sydney.

LEE, E.S. (1966) 'A theory of migration', *Demography* 3, pp. 47–57.

LUCKMAN, L., GIBSON, C., FITZPARTICK, D., BRENNAN-HORLEY, C., WILLOUGHBY-SMITH, J. & HUGHES, K. (2009) *Creative Tropical City: mapping Darwin's creative industries*, Charles Darwin University, Darwin.

MARKUSEN, A. & GADWA, A. (2009) 'Arts and culture in urban/regional planning: a review and research agenda', Working Paper 271, project on regional and industrial economics, University of Minnesota, Minneapolis.

MARKUSEN, A., GILMORE, S., JOHNSON, A., LEVI, T. & MARTINEZ, A. (2006) 'Crossover: how artists build careers across commercial, nonprofit and community work', project on regional and industrial economics, Humphrey Institute of Public Affairs, University of Minnesota, Minneapolis.

MOORE, H. & HELD, D. (eds) (2007) *Cultural politics in a global age*, One World, Oxford.

O'NEIL, H. (2009) 'Ratbags at the gates', in Schultz, J. (ed.) *Essentially creative*, *Griffith Review Autumn 2009*, ABC Books, Sydney, pp. 11–40.

PECK, J. (2005) 'Struggling with the creative class', *International Journal of Urban and Regional Research* 29(4), pp. 740–70.

PRATT, A. (1997) 'The cultural industries production system: a case study of employment change in Britain, 1984–91', *Environment and Planning A* 29(11), pp. 1953–74.

PRATT, A. (2009) 'Urban regeneration: from the arts "feel good" factor to the cultural economy: a case study of Hoxton, London', *Urban Studies* 46(5–6), pp. 1041–61.

RIZVI, F. (2000) 'International education and the production of the global imagination', in Burbules, N. & Torres, C. (eds) *Globalization and education: critical perspectives*, Routledge, New York, pp. 205–27.

STRATTON, J. (2008) 'The difference of the Perth music: a scene in cultural and historical context', *Continuum* 22(5), pp. 613–22.

THROSBY, D. (2008a) 'Creative Australia: the arts and culture in Australian work and leisure', Occasional Paper 3/2008, Academy of the Social Sciences in Australia, Canberra.

THROSBY, D. (2008b) 'The concentric circles model of the cultural industries', *Cultural Trends* 17(3), pp. 147–64.

THROSBY, D. & HOLLISTER, V. (2003) *Don't give up your day job: an economic study of professional artists in Australia*, Catalogue No. 331.7617, Australia Council, Sydney.

VANOLO, A. (2008) 'The image of the creative city: some reflections on urban branding in Turin', *Cities* 25, pp. 370–82.

Creative Migration? The attraction and retention of the 'creative class' in Launceston, Tasmania

MADELEINE VERDICH, *WorleyParsons Pty Ltd[1], Brisbane, Australia*

ABSTRACT   *Concepts of creativity and the attraction of a 'creative class' have become increasingly prominent in regional economic development literature and policy. Richard Florida's books 'Rise of the creative class' and 'Who's your city?' have encouraged city and regional planners to move away from strategies focused on infrastructure development and the attraction of businesses towards strategies which attract people as migrants, particularly the 'creative class', through a focus on characteristics such as a 24/7 lifestyle, cultural amenity and ethnic diversity. This research explored why people who could be categorised as 'creative class' move to Launceston in Tasmania, and what keeps them there. In the regional centre of Launceston, lifestyle, amenity and diversity were not characteristics that attracted in-migration initially. Instead, characteristics particular to small and rural places attracted creative and other professional workers, such as outdoor amenities, downshifting, time with family, proximity to the natural environment and a strong sense of community. Only after arrival did Launceston's comparative cultural wealth come to be appreciated by in-migrants, instead becoming a factor helping to retain newly arrived migrants.*

Introduction

The last decade has seen an explosion of public debate about encouraging cities and regions to become more 'creative'. This drive towards creativity has been associated with the increasingly central role of city competitiveness in regional development and planning. Sitting squarely the middle of this debate is the American academic Richard Florida, whose creative capital theory has grasped the attention of city and regional planners and economic development practitioners throughout the world. Florida's theory assumes that the economy is increasingly a 'creative economy' driven by a 'creative class' (a social group of professional workers who work in jobs that require innovation and creativity, but who also consume cultural products avidly), so that for cities and regions to succeed, they must attract members of this creative class (Florida 2002). According to Florida, 'the question that lies at the heart of our age' and is central to the competitiveness of towns and cities in the new economy, is: '*How do we decide where to live and work?*

129

What really matters to us in making this kind of life decision? How has this changed and why?' (2002, p. 217; emphasis in original).

In an era of economic restructuring and the diminishing importance of traditional manufacturing and agricultural industries, new and innovative solutions are sought and service industries developed to maintain the viability of regional centres. Policy makers have consequently begun to look at the potential of social capital, human capital and more recently creative capital to encourage economic growth and attract new residents, particularly those who can contribute to the diversification and development of the local economy. Florida's thesis identifying the factors which drive creative-class migration has begun to attract interest in the Australian regional economic development policy arena (National Economics 2002; DOTARS 2003; Creative Class Group 2009) as has his claim that cities and regions must create attractive environments in response to these drivers. Yet despite the urban-centric nature of Florida's theory, and the academic criticism it has received (Peck 2005; Gibson & Klocker 2005; Glaeser 2004), little, if any, locality-specific qualitative or quantitative research has been undertaken to assess the validity of his thesis in a regional context, and examine why creative people might move to and stay in regional centres. Indeed, current research lacks analysis of the distinction between the attraction and retention of creative-class migrants (see Florida 2002; National Economics 2002; Keniry *et al.* 2003), thus assuming that the factors which motivate people to move to cities and regions are the same as those that encourage them to stay. This article goes some way to bridging this gap by using empirical evidence from regional Australia—Launceston, Tasmania—to trace what factors drive the attraction and retention of members of the creative class to regional areas, and the extent to which these factors support Florida's creative capital theory.

Attracting the creative class: the great debate

A core argument emanating from the creative capital thesis is that in order to encourage and sustain economic growth, cities and regions must attract the creative class. Indeed, the essence of the theory states that, 'regional economic growth is driven by the location choices of creative people—the holders of creative capital—who prefer places that are diverse, tolerant and open to new ideas' (Florida 2002, p. 223). Thus, as Baris (2003, p. 42) contends, 'the old mode of people moving to follow jobs is turned on its head' and in order to compete in the new race for talent, cities must 'restructure themselves for the Creative Class's needs'. A key component of this, and one which is central to this article, concerns the importance of 'quality of place' in the attraction and retention of the creative class. Human and creative capital theories assert that economic growth will occur in places that have highly educated people; further questions are then raised about the reasons why some regions are more attractive than others for this group of people. The factors that dominate creative-class locational decisions are regarded by Florida as 'so powerful ... I have coined a term to sum them up: *quality of place'* (2002, p. 231; emphasis in original). This term, in contrast to the more traditional concept of quality of life, refers to 'the unique set of characteristics that define a place and make it attractive' (Florida 2002, p. 231). Subsequently these have been elaborated as a combination of economic opportunity, services, leadership (business and political), values (notably tolerance and trust), aesthetics and lifestyle, though a different balance would influence different people (Florida 2008).

Whilst developing his creative capital model, Florida encountered six consistent themes which can be seen as the 'hallmarks' of locations that are attractive to the creative class (Prins 2005). These hallmarks of attractive locations are a '24/7' lifestyle; diversity; authenticity; identity; opportunities for social inclusion; and thick labour markets. The emphasis of Florida's hallmarks is on openness and the ability for people to be themselves. Ultimately, Florida argues, it was neither the amenities nor the bohemians or gay people that accounted for the attraction of creative people, but rather the 'low barriers to entry' for human capital (Florida 2002). Consequently, diversity of people, authenticity of place and the ability to engage in social interactions have been seen as central components of those cities which have been successful in attracting creative and skilled people.

Florida's theory of creative capital is undeniably urban-centric, even metropolitan-centric, and arguably displays little interest in the prospects of regional areas, which appear to have far less to offer to the mobile 'creative' workforce (Prins 2005). Florida acknowledges this urban bias in the creative-class model. Indeed, he expresses concern that the social shifts he has observed are creating a new social divide, stating that:

> I fear we may be splitting into two distinct societies with different institutions, different economies, different incomes, ethnic and racial make-ups, social organizations, religious orientations and politics. One is creative and diverse—a cosmopolitan admixture of high-tech people, bohemians, scientists and engineers, the media and the professions. The other is a more close-knit, church-based, older civic society of working people and rural dwellers. (2002, p. 281)

Some within Australia (Prins 2005, p. 1) have taken this comment as a warning for places other than big cities, arguing that it presents a 'bleak picture of how regional economic disparity is reinforced through social patterns, such as migration . . . and the inexorable "brain drain" of skilled people to larger cities'. In contrast, Sorenson (2009) argues that rural and regional areas and the people who live there display the fundamental qualities of creativity. This article seeks to extend this debate. I argue that far from being part of an 'older civic society of working people' (Florida 2002, p. 281), those I interviewed in Launceston appreciated and were involved in and with creative and cultural events, facilities and a milieu—yet, crucially, the motivations which attracted them to Launceston were rather different from those identified by Florida as the key drivers of creative-class migration. Involvement in cultural and creative activities followed in-migration, but did not determine it.

The Launceston story

Launceston is the second largest city in Tasmania and home to approximately 62 000 residents. The city is the major service centre for the north of the State and has an extensive European and Indigenous history, with 200 years of European settlement and 35 000 years of Tasmanian Aboriginal presence. Although air travel has become more affordable and frequent over the last few years, Launceston, and indeed Tasmania more broadly, has remained relatively removed from the global metropolises of the 'mainland': the only Australian State without international air connections. Nonetheless, Launceston does have a relatively high number of the cultural and lifestyle assets which Florida (2002) has claimed as essential drawcards for the creative class.

The population has remained relatively stable over the last 10 years, although in an outcome which bucked the broader Tasmanian trend, Launceston achieved modest population growth between 2001 and 2006 (ABS 2006). Demographically, Launceston is broadly characteristic of rural and regional Australia, with low levels of ethnic diversity and an ageing population. There are some variations, including an increase in humanitarian visa holders from Sudan and Sierra Leone settling in Launceston in the past 5 years. Unlike many regional areas which have experienced the 'brain drain' phenomenon (Gabriel 2002) of young people leaving to gain experiences elsewhere, the Launceston demographic profile indicates a proportionately greater number of 15–24 year olds than Tasmania or Australia more broadly. This 'bulge' probably reflects the presence of a university and matriculation-aged population (Jackson 2005), with anecdotal evidence gathered through fieldwork for this paper suggesting that though some young people move away for university and other opportunities after school, others move into Launceston to attend the university.

Launceston has a strong industrial history, and indeed manufacturing continues to be a key presence, albeit waning (Bugg 2005). In the past decade, manufacturing experienced negative employment growth (–4 per cent) while property and business services increased by 87 per cent and personal and other services by 68 per cent. Launceston has thus moved from being regarded as a vulnerable manufacturing base in 2003 to an advantaged service-based city in 2006 (Baum 2003, 2006).

Launceston has a depth of cultural, educational and natural assets. The city is host to tertiary education institutions including the University of Tasmania, the Australian School of Fine Furniture and cultural infrastructure such as the Queen Victoria Museum and Art Gallery (QVMAG), established in 1891. The University of Tasmania's Launceston Art Academy, Art Gallery and School of Architecture is collocated with QVMAG's Inveresk site in the city's converted rail workshops. A diverse mix of good-quality restaurants and cafés have arrived recently, inspired by high-quality niche Tasmanian agricultural and viticultural products. The city, located at the junction of the North Esk, South Esk and Tamar Rivers, is proximate to a range of recreational opportunities and natural amenities such as Cataract Gorge which provides opportunities for swimming and hiking and Mt Stromlo, the State's only ski field. This fabric of cultural institutions and amenities has played an important role in the development of the city's identity and the further development of a unique product and service base now evident in Launceston. For at least some interviewees in this research, this depth of cultural infrastructure has played an important role in Launceston's maturity to a more vibrant city.

Methods

In light of Florida's creative capital hypothesis for drivers of migration, one might expect that the city's cultural fabric, quality restaurants and access to tertiary institutions would figure heavily with the values and activities of recent migrants. The objective was therefore to understand the importance of culture and creativity in the attraction and retention of this skilled professional group. Appreciation of the factors attracting and repelling creative professionals enables an evaluation of

Florida's theory in relation to migrant experiences outside major cities. In-depth semi-structured interviews were conducted with 18 migrants in Launceston, in order to ascertain whether cultural activities, creativity and amenity were factors underpinning migration. Interviewees were initially sourced through the use of promotional material distributed through 'gatekeeper' organisations and later using a snowballing technique. All of the interviewees fit within, or were in a relationship with someone who fit within, Florida's definition of the 'creative class'. That is, they were 'Creative Professionals', the classic knowledge-based workers including those working in healthcare, business and finance, the legal sector, and education and the 'Super-Creative Core', defined as scientists, engineers, innovators, and researchers, as well as artists, designers, writers and musicians. Although their average age was 27, 10 were in their 30s, all but two were in steady relationships and six had children. Although occupationally and recreationally 'creative class', the sample stood in some contrast to the stereotypical young, hip and single image of the creative class that Florida portrays. The qualitative and semi-structured nature of the interviews produced a series of narratives used to evaluate how individuals perceive their migration experiences and the place in which they live.

The 'attraction factor'

How, then, did creative people get to Launceston? Only once was Launceston chosen specifically because of the cultural and recreational amenities provided and its 'cosmopolitanism'. Respondent S and her partner work in the mining and resource sector and moved to Launceston 6 months prior to being interviewed. S is a Senior Environment Scientist with highly sought after skills and had a range of job opportunities throughout Australia. Although this provided security, the majority of jobs were located in remote mining communities with limited access to cultural services and amenities. After 10 years spent in such communities, migrating to Launceston allowed them to access the culture, creativity and amenity that existed in a larger service centre. S further reiterated that their main aim was to 'be able to live in a community that had good quality restaurants, cafés, a cinema and a host of other cultural activities and facilities'. Size and scale thus play a part.

In contrast to this lone case, for the other interviewees, migration to Launceston was characterised predominantly by the wish for a better quality of life or lifestyle—with quality of life and lifestyle conceptualised differently from Florida's more urban-centric ideas. The terms quality of life and lifestyle were used interchangeably and understood as a single notion pegged to the simple idea that a small place, close to rural hinterland and wilderness, offered something tangibly different from city life (Bell & Jayne 2006). Migration dominated by lifestyle factors was driven by the wish to enjoy a quality of life which they considered unobtainable in larger metropolitan cities. Tasmania, and more specifically Launceston, provided a lifestyle that combined proximity to the natural environment, a slower pace of life and the fresh air of the country. This 'clean' and 'wholesome' lifestyle had become rare or extinct in the 'big city' owing to the constraints of traffic, time and scale.

The desire to create a more balanced life that emphasised family time and participating in a range of non-work activities, rather than money and social position, was brought up by a number of people—evidence of the increasingly popular 'downshifting' phenomenon (Hamilton & Mail 2003). Prior to arriving in

Tasmania six participants had been engaged in highly stressful work environments in large cities. None of them wanted similar employment or remuneration, but preferred that their quality of life improved. This sentiment was expressed by Respondent H who arrived in Tasmania in 2004, 2 years prior to the study, with his wife and newborn child:

> I was working for a multi-national company covering the Asia-Pacific involved in learning development. I was burnt out and travelling a lot so I wanted to focus on a better work life balance and I did an internet search and Launceston came up. I bought a house in rural Tasmania. I moved here, one for the property prices and two, to have a better work–life balance—I didn't care what I did—I didn't care if I was licking stamps for a job.

As the major metropolitan cities grow in size, problems such as increasing congestion, noise, air pollution, travel times, crimes and other 'diseconomies of scale' are contributing to the flight of many of these individuals to more liveable places such as Launceston (Bell 1996, p. 15). Congestion and the waste of valuable time were raised as key push factors when deciding to leave the city. For Respondent R:

> The day I chose to move back here was the day I was sitting at the traffic lights in Melbourne thinking I spend 27 hours a week in my car, this is not the way I want to spend this much of my life. I can't get over how many people who say they just can't get over the lack of privacy, the time wasted in transport, pollution, and then crime and things are further down, it's this waste of time on non valuable things.

The diversity and magnificence of the natural environment present in Tasmania is the backbone of State and local tourist campaigns. Similarly, the proximity of Launceston to the Tamar Valley and River, Cataract Gorge and the coast were an important consideration for people wishing to move to Tasmania. The Executive Officer of Launceston Chamber of Commerce commented that she was amazed by the number of enquires that the chamber received from young people who were motivated to move to Launceston for the 'wilderness and being near walking and diving and outdoor things'. As respondent P said after returning from Auckland: 'There I felt I was driving round with a sign on the back of my head saying country-girl—it was all so hectic and busy, I was relieved when I returned to Launceston.' In North America, too, the natural environment plays a part in attracting creative-class members to rural and regional areas: as in Launceston, 'the appeal of natural amenities and associated recreational opportunities is sufficiently strong for many in the creative class to locate in rural areas rich in outdoor amenities and that this movement is associated with rural growth in employment and population migration' (Wojan & McGranahan 2007, p. 198).

Family ties were important. Of the 18 people interviewed, only two were single (both of whom were under the age of 25), six participants had children and two were looking to have children in the near future. The aspiration to bring children up closer to one's immediate family or in a healthier environment than could be found elsewhere featured prominently. Several respondents cited the desire to live with their partner as the prime reason they migrated. Upon having children, there are strong motivations to be closer to one's immediate family for support as well as enabling family members to participate in childcare and upbringing. Respondent G moved to Launceston from Canada with her partner and their young child in 2002, 4 years prior to the study. G's partner had grown up in Launceston prior to migrating to Canada and their return was a combination of 'moving home' and being closer to her parents-in-law who had also returned to Tasmania. The move not only provided them with a 'new start' but also allowed the child's grandparents to be involved in his life. This proximity to their family was not only a prime reason attracting them to Launceston but was central to their idea of quality of life in a way which differs markedly to the way quality of life has been understood in big cities ('24/7 lifestyle' etc.). The desire to be closer to families and partners was a prominent theme established by those interviewed, which shows how the same family factor that underpin much international migration (Boyd 1989) also influence internal migration.

Economic motivations for migration were relevant for three of the interviewees. However, in all but one case, respondents stated that although they moved to Launceston as a result of gaining employment, the acquisition of a better quality of life or family commitments was the deciding factor. Migration was rarely a purely economic decision based on salary or better job opportunities, contrary to many of the econometric studies of migration (e.g. Clarke 1982; Stark & Taylor 1991) which assume that people will act as rational economic beings and move wherever the market needs them. Downshifting challenges such narrow 'rationality'.

This was well illustrated through the experience of two respondents. C and her husband migrated from South Africa 9 months prior to the interview. After her husband's contact with the mining company BHP Billiton ended, he was given the option of relocating to Mexico or Tasmania, or stay in South Africa and be retrenched. Although they believed they would be able to maintain a higher standard of living in both Mexico and South Africa than Tasmania, the quality of life attainable in Tasmania would be considerably higher. Thus, although the reason to migrate to Launceston was ultimately related to employment, in that without a job it would never have been considered, the decision to choose Tasmania over South Africa or Mexico was based on lifestyle.

More evidently rural objectives were important for some. An interest in permaculture and a desire to build their own home led K and L to become interested in the Bachelor of Environmental Design course at the Launceston University campus. Emigrants from England, they stated that 'Launceston was chosen out of all the other places in Australia because of the University'. K believed that not only would he be able to enrol on the course but that he would also be able to gain employment at the university through his experience in administration. The move to Australia was motivated by the wish for a lifestyle change, yet the choice of Launceston specifically was based on the presence of the university and the unique course that it offered.

Overall, cultural factors were rarely associated with quality of place as the key or even a secondary reason for migration among this group of skilled professionals. Family ties and time, environmental concerns and a more peaceful existence were rather more important, even though cultural attractions were accessible.

Finding 'home'

Although Florida (2002) emphasises the highly mobile and even 'flighty' nature of the creative class, migration is not a simple or speedy process and, as a result, rarely do people simply pack up and leave as soon as they have previously moved, particularly those with children. Often the reasons to stay in a place for a short to medium period of time mirror those which motivated the initial move: as long as prior expectations are realised the location becomes and remains an attractive place in which to live. This research explored not just the motivations for professionals to move to Launceston but also the things that kept them there. In other words, what features of a place contribute to deeper commitments by migrants to make a place their 'home'? The reasons for individuals or families remaining in a place might therefore differ from those which influenced their decision to migrate.

In Launceston, quality of place and the cultural and cosmopolitan amenities associated with creative industries became markedly more important when individuals were deciding to make a place their home, so envisaging staying there. To many participants, cultural facilities and amenities in Launceston were a pleasant surprise, adding to a sense of the city life, and surpassing expectations. Because of these cultural amenities, Launceston was more likely to encourage the retention of skilled migrants after their initial move. For those respondents who felt the services and amenities available in Launceston exceeded their expectations, all but two believed they would remain in the city long term and make it 'home', even if that had not been their initial aim:

> The vibrancy and cultural heritage of Launceston is on the cusp of being a motivating factor to stay and it has been a bonus. (Respondent F)

The above statement epitomises the feeling portrayed by the majority of respondents with regard to the role that cultural and cosmopolitan facilities and amenities play in the decision to stay in Launceston. The high-quality food and wine scene present in Launceston added to people's quality of life and their enjoyment of the area. It added vibrancy to the city and the surrounding region and there appeared to be little concern among respondents that they were 'missing out on something'. This food and wine scene has further enhanced the enjoyment of place.

Despite the considerable number of participants who stated that cultural facilities and amenities were important factors in deciding to stay, very few had much contact with the physical cultural infrastructure. Thus the majority of participants had only attended the QVMAG once or twice and perceived it more as something that you sent visiting relatives and friends to. Those facilities which were more highly frequented and valued were restaurants, bars and cafés, the theatre, symphony orchestra events, parks and recreation areas and Festivale, the annual food and wine festival. These venues provide for what Florida terms 'opportunities for social interaction' (2002, p. 226).

Ultimately, however, a number of factors surpassed cultural facilities and amenities in respondents' decisions to remain in Launceston and identify it as home. The most prominent narratives were those of lifestyle and family. For some respondents, regardless of whether they left for a short period, they would always come 'home' to Launceston because that was where their family was and their roots were. For others, the decision to stay hinged upon the future employment opportunities for their partner, where a small city might not have enough choices. Lifestyle remained an important consideration in decisions to stay for many respondents, provided they could maintain the lifestyle they had obtained upon arrival.

Creative migration—does Launceston 'get it'?

The creative capital thesis suggests that the decision to move to, within or between cities is associated with the freedom of anonymity, superior career and educational opportunities and the desire to engage in diverse lifestyles (Florida 2002). Locations must facilitate and encompass diversity, the presence of desired scenes, access to creative pursuits and the provision of amenities that accommodate the 'experiential life' which is said to be so vital to the creative class. Yet in the much smaller city of Launceston this is not true. The findings of this research dispute this interpretation, in a regional context where large cities were perceived as oppressive rather than centres of life and creativity. Links between quality-of-life considerations and migration to Launceston were explicitly underpinned by images of place that had little to do with Florida's themes.

The lifestyle demands of the creative class as claimed by Florida (2002) vary significantly from those expressed by the members of the creative class in Launceston. Although Florida does, to some extent, refer to the importance of the proximity of the natural environment, more emphasis is placed of the presence of 'scenes', such as the music scene, arts scene, technology scene and a vibrant nightlife (Florida 2002, p. 224). He contends that 'time and time again, the people I speak with say that these things [...] signal that a place "gets it"—that it embraces the culture of the Creative Age'. Lifestyle requirements outlined by those migrating to Launceston from metropolitan areas focused not on the ability to access 'scenes' or 'on demand entertainment' (Florida 2002, pp. 224–5) but rather revolved around proximity to the natural environment, a less stressful existence and a safe and healthy environment for their children. The long hours at work, which Florida argued were so fundamental to the creative class, were, in contrast, 'push factors' for participants, albeit older than those depicted by Florida. Indeed, many sought refuge in Launceston from their stressful and time-poor city lives; as P stated: 'I really wanted and needed a break from my career and it got to a point where I needed to physically remove myself from the environment that I was in.'

These lifestyle imperatives are consistent with the findings of a study of migration to sea-change localities (e.g. Burnley & Murphy 2004), though many such migrants were on the verge of retirement. Rather than using the term lifestyle, the study refers to 'better environment' which includes the 'desire for peace and quiet, security needs, and less crime, as well as aesthetics of the natural environment'.

The rise of the creative class contends that 'movement away from communities of strong ties to communities of weak ties' is the new trend in modern life and contests the validity of social capital for economic development (Florida 2002, p. 261). Yet this concept that community is no longer desired by people is in striking contrast to the narratives expressed in Launceston. Far from the desire for 'quasi-anonymity', participants sought to *escape* the anonymity of large metropolitan areas for places with a strong sense of community. This was evident not only in discussions concerning reasons for migration but also in perceptions of which cultural amenities were most important for quality of life.

Community-oriented events such as Festivale and the Tasmanian Symphony Orchestra's 'Music in the Park' were amenities that were both well attended and played a vital role in the city's identity and quality of place. Similarly, narratives of a desire to belong to a community and make a tangible contribution to it were expressed by many respondents. In Tasmania, individuals have the ability to be 'a big fish in a small pond' (Respondent E) and consequently there are more opportunities to be involved in the governance and development of localities, cities and/or the State: more opportunities for creativity. It is important to recognise, however, that some respondents had experienced elements of this community cohesiveness as entry barriers rather than pull factors. Central to this was a perception that the community was highly parochial and conservative, which some participants believed had affected their ability to gain employment or introduce new, innovative systems or processes into their places of employment. The distinction made between 'mainlanders' and 'locals', which some Tasmanian respondents suggested strengthened social capital and a sense of community, has been an isolating rather than inclusive experience for some respondents. The labelling of people as 'different' had the potential to instigate high social entry barriers and indeed went against expectations of a close and inviting community, central to the rural and regional idyll. Even so, the respondents sought more rather than less community spirit.

So does Launceston 'get it'? According to Florida, a city 'gets it' if it is able to adapt to the demands of the creative age, and create an "environment or habit that is attractive to the creative class" (2002, p. 302). The question as to whether Launceston 'gets it' refers to its ability to create an environment which provides for the wants and needs of *its* creative class. While Launceston had done this, the creative migrants who have moved there were distinctly more mature and family oriented than those described by Florida. In order to continue to attract individuals who through their creativity are able to contribute to the local economic and social fabric, Launceston's policy makers and planners must further encourage community inclusiveness, innovation and openness and create an environment which is able to sustain the lifestyle opportunities currently attracting members of the creative class—even though that creative class is not that described by Florida.

Conclusions

Florida's creative-class migration theories have played an important role in encouraging new ways of conceptualising and approaching metropolitan economic development, and of diversifying and enhancing creativity. For governments searching for new responses to the increasing polarisation between those regions that are winners and those that are losers, creativity and cultural planning present an attractive and seemingly uncontroversial strategy (Gibson & Klocker 2005). Indeed, if one goes by the list of those towns identified by Florida, including Noosa on Queensland's Sunshine Coast, which are said to be improving their economic and social situations as a result of incorporating these theories into their policy and planning, it could be assumed that this is a winning recipe (www.creativeclass.org). However, it is clear that, at least in the case of Launceston, the creative migrants who are moving and staying there are quite different from those purported to be at the cutting edge of urban creativity, and their reasons for moving are rather more prosaic and humble.

This study provides an alternative view of the factors which drive creative-class migration to rural and regional areas. Importantly, it provides evidence that the concept of 'quality of life' is often understood differently by those residing in a small, regional city than it is by big-city dwellers. Smallness influences how places are imagined and experienced (Waitt & Gibson 2009; Bell & Jayne 2006) and plays an important role in individuals' perception of quality of life. In the context of Launceston, its small scale is perceived as a safe haven to escape the rat race of the city, thus feeding into desires to move there. This is in striking contrast to the presence of cultural facilities and amenities which, whilst present in Launceston, did not play a role in the decision to migrate for the vast majority of those interviewed. That is not to say, however, that such amenities do not influence perceptions of quality of life. Rather, enjoyment of cultural facilities and amenities featured later in people's narratives of migration experiences—becoming something of a retention factor rather than a marketable quality to attract professional migrants in the first place.

This research illustrates the importance of understanding the attributes of specific locations in order to develop appropriate and relevant economic development and promotion strategies, rather than taking a 'one-size-fits-all' approach. Florida's assertion that cities and towns must provide multiple 'scenes', a 24/7 lifestyle and ethnic and sexual diversity in order to survive proved to be less pertinent. In Launceston and elsewhere (Wojan & McGranahan 2007; Rockhampton Regional Development Limited 2007) particular rural and small-scale characteristics tend to attract workers in creative occupations (and others) and it is these rural qualities, rather than urban-centric 'scenes', which should be maintained and further developed to encourage the migration of the creative class out of the metropolis. It may even be that the converse—developing such characteristics in large cities and metropolitan centres—might be at least as beneficial as developing those espoused Florida.

NOTE

[1] The views expressed here are those of the author, and may not be attributed to WorleyParsons Pty Ltd.

REFERENCES

AUSTRALIAN BUREAU OF STATISTICS (ABS) (2006) *Times series profile*, Catalogue No. 2003.0, AGPS, Canberra.

BARIS, M. (2003) 'Book review: The Rise of the Creative Class', The Next American City 1, pp. 42–5.

BAUM, S. (2006) 'A typology of socio-economic advantage and disadvantage in Australia's large non-metropolitan cities, towns and regions', *Australian Geographer* 37(2), pp. 233–58.

BELL, D. & JAYNE, M. (2006) 'Conceptualising small cities', in Bell, D. & Jayne, M. (eds) *Small cities: urban experience beyond the metropolis*, Routledge, London, pp. 1–18.

BELL, M. (1996) *Understanding internal migration*, AGPS, Canberra.

BOYD, M. (1989) 'Family and personal networks in international migration: recent developments and new agendas', *International Migration Review* 23(3), pp. 638–70.

BUGG, P. (ed.)(2005) *Northern Tasmania economic bulletin*, Northern Tasmanian Development Board, July.

BURNLEY, I. & MURPHY, P. (2004) *Sea change: movement from metropolitan to arcadian Australia*, UNSW Press, Sydney.

CLARKE, G. (1982) 'Dynamics of inter-state migration', *Annals of the Association of American Geographers* 72, pp. 297–313.

CREATIVE CLASS GROUP (2009) *Noosa: Creative Communities Leadership Program*, Noosa Creative Alliance, available from: www.noosacreativealliance.com.au (accessed 2 October 2009).

DOTARS, R. (2002) *The rise of the creative class: and how it's transforming work, leisure, community, and everyday life*, Basic Books, New York.

FLORIDA, R. (2008) *Who's your city?*, Basic Books, New York.

GABRIEL, M.A. (2002) 'Australia's regional youth exodus', *Journal of Rural Studies* 18, pp. 209–12.

GIBSON, C. & KLOCKER, N. (2005) 'The "cultural turn" in Australian regional economic development discourse: neoliberalising creativity?', *Geographical Research* 43(1), pp. 93–102.

GLAESER, E. (2005) 'Review of Richard Florida's The Rise of the Creative Class', *Regional Science and Urban Economics* 35, pp. 593–6.

HAMILTON, C. & MAIL, E. (2003) 'Downshifting in Australia: a sea-change in the pursuit of happiness', Australia Institute Discussion Paper No. 50, January.

JACKSON, N. (2005) 'Tasmania's population', available from: www.taspop.tasbis.com (accessed 9 June 2006).

KENIRY, J. *et al.* (2003) *Regional business: a plan for action*, Commonwealth Department of Transport and Regional Services, Canberra.

NATIONAL ECONOMICS (2002) *State of the regions report 2002*, Australian Local Government Association, Clifton Hill, Vic.

PECK, J. (2005) 'Struggling with the creative class', *International Journal of Urban and Regional Research* 29(4), pp. 740–70.

PRINS, S. (2005) 'Rhetoric or reality in the new Tasmania', Evatt Foundation, available from: http://evatt.labor.net.au (accessed 9 September 2006).

ROCKHAMPTON REGIONAL DEVELOPMENT LIMITED (2007) *Central Queensland of opportunity*, Rockhampton Regional Development Limited, Rockhampton.

STARK, O. & TAYLOR, J. (1991) 'Migration incentives, migration types: the role of relative deprivation', *The Economic Journal* 101, pp. 1163–78.

WAITT, G. & GIBSON, C. (2009) 'Creative small cities: rethinking the creative economy in place', *Urban Studies* 46(5&6), pp. 1223–46.

WOJAN, T. & McGRANAHAN, D. (2007) 'Recasting the creative class to examine growth processes in rural and urban counties', *Regional Studies* 41, pp. 197–216.

Indigenous Hip-hop: overcoming marginality, encountering constraints

ANDREW WARREN & ROB EVITT, *University of Wollongong, New South Wales, Australia*

ABSTRACT   *This paper discusses the creative and contemporary performances of young Indigenous hip-hoppers in two seemingly disparate places (Nowra, NSW, and Torres Strait Islands, QLD). Visiting two Indigenous hip-hop groups from these places—and drawing on interviews and participant observation—we explore the way in which emerging technologies, festivals, programs and online networking have helped enable unique forms of music making. In contrast to racist discourses depicting Indigenous youth as idle or inactive, our research participants demonstrated musical aspiration, creativity and a desire to express love of country and culture. Rather than assume cities and urban centres are hubs for creativity, hip-hop production is geographically mobile, operating in locations removed from large population centres. Indigenous hip-hop links up-and-coming with more experienced performers in what amounts to a semi-formal, political, transnational and anti-colonial creative industry. Geographical distance remains an ongoing challenge, but more than this, wider patron discourses framing what is expected from 'proper' Indigenous performance are the more profound coalface of marginalisation.*

Introduction

This paper explores the creative musical performances of Indigenous youth from two socio-economically disadvantaged places—one in Australia's tropical north, the other just beyond the outermost edge of the Greater Sydney metropolitan area. In these locations, physical distance and poverty are conditions influencing the ability of creative artists to do their work, access opportunities and build careers. We discuss these themes in relation to young Indigenous people involved in the musical genre of hip-hop. We discuss how remoteness is managed, and marginality negotiated through the expressive medium of hip-hop and new recording and distribution technologies. In doing so, we seek to explore a network—semi-informal, political, transnational and often decidedly anti-colonial—which constitutes a new, vernacular, Indigenous creative industry in regional and remote Australia.

But crucially, we also explore how physical distance and poverty are not the only barriers that creative artists negotiate. Young musicians navigate expectations of themselves and what constitutes 'proper' Indigenous performances in wider Australian cultural industries. We draw on van Toorn's (1990) concept of patron discourses to show how, beyond physical and socio-economic marginality, cultural norms and expectations frame what is possible, producing and restricting creative opportunities.

This article is a collaboration between two researchers—one Indigenous, one non-Indigenous (both having grown up in the Southern Illawarra)—who brought to this project different goals and backgrounds. For one this work contributes to a PhD thesis on young people as cultural assets in regional Australia (funded through the ARC Linkage Project Cultural Asset Mapping for Regional Australia—see http://culturemap.org.au/). The other is both a student and active member of the region's Indigenous hip-hop scene.[1] This collaboration provided unique links and personal connections through which fieldwork could be pursued (see below) but also constituted an example of research practice heeding calls in Aboriginal studies literature to bring to the fore critical questions of research ethics, subjectivity and Aboriginality (Langton 1993; Gibson 2006; Burarrwanga et al. 2008). It is as much an article born of a dialogic conversation between two researchers as it is the outcome of a regular research project.

Hip-hop: glocal, transnational, mobile music

Originating as a music format in the disadvantaged urban neighbourhoods of the Bronx, Harlem and Brooklyn in New York during the 1970s (Kitwana 2003), and further building on Jamaican sound system culture (Bradley 2000), hip-hop traditionally involves dee-jaying (beat), rapping (MC), break dancing (B-boying) and graffiti elements. Its commercialisation has steadily transformed the genre from underground element of urban culture to mainstream, global industry, with its own distinct language (Samy Alim 2007), spawning fashion brands FUBU and Tommy Hilfiger. Aspiring young rappers have begun replacing dee-jaying elements with music software allowing at-home creation, cutting together different instruments to form unique instrumental sounds or beats.

A distinct discourse of locality and authenticity surrounds hip-hop (Pennycook 2007). For many black American youth in the 1980s and 1990s, disenfranchised with life in urban ghettos, hip-hop enabled articulation of oral stories confronting daily life on the streets; gang-related violence, extortion and drug dealing. As a way of traversing, navigating and making sense of daily struggles, youth turned to hip-hop for enlightenment (Kitwana 2003). Many listened and danced to rappers and DJs playing on street corners, before trying hip-hop. As underground and oppositional, street-performed hip-hop grew increasingly popular, drawing large crowds for neighbourhood 'bloc parties' (Toop 1984).

By the late 1990s, hip-hop was the highest-selling musical genre in America, with sales of 81 million albums in 1998 alone (Farley et al. 1999). Commercial, yet still confrontational and oppositional, hip-hop was globalised via CD, television and fashion, becoming a vehicle of expression and identification, particularly relevant for working-class, migrant and Indigenous youth (Mitchell 2003, 2006). Disenfranchised groups related to the identities and circumstances behind the music; it was linguistically powerful, at times arrogant—a platform where minority bodies and voices were thrust into hegemonic and vice-regal positions in the media landscape (especially in music video clips).

In Australia hip-hop became popular amongst Indigenous youth, where influence from American hip-hoppers Ice Cube, TuPac, Snoop Dogg and Jay Z was strong. The uptake of hip-hop by Indigenous Australia can also be attributed to an evolving 'transnational black culture' (Dunbar-Hall & Gibson 2004; White 2009). While hip-hop is global language, positioned around ideas of brotherhood and resistance, it is also an open soundtrack for interpretation and 'flushing' by local experiences, for 'the articulation of Aboriginal identity based on the valorisation of blackness' (White 2009, p. 108).

When performing in Australia, artists such as Snoop Dogg and Ice Cube have made efforts to connect with local Aboriginal populations, referring to cultural similarities during interviews and gigs, while making physical contact with communities, as Snoop Dogg recently did in Redfern in inner-city Sydney. A number of local performers have also played a significant role in indigenising hip-hop in Australia, actively tutoring and mentoring emerging and grassroots enthusiasts, hosting workshops, teaching skills and providing direct musical support. Aboriginal performers MC Munki Mark, Wire MC and Brotha Black provide cultural and creative learning for budding Indigenous artists (Mitchell 2006). Their music advocates pride and solidarity, projecting Indigeneity as brotherhood. These pedagogies help develop slick rhyming and language, performance and self-confidence through rapping.

Not restricted to Sydney, Melbourne or Brisbane, Indigenous musicians are practising and performing hip-hop in remote and isolated communities in places such as Wilcannia (Wilcannia Mob), Bowraville (Bowra Rhythm Mob) and Kununurra (G-Unit). These locations have burgeoning hip-hop scenes despite geographical distance from Australia's traditional music industry centres. Despite emergent remote creative scenes, Indigenous hip-hop outside capital cities is yet to be documented or analysed academically. Where Indigenous popular music from remote areas has been a feature of academic work (e.g. Gibson & Connell 2004; Dunbar-Hall & Gibson 2004), it has often focused on country, reggae and rock styles. This article redresses that lacuna, drawing attention to Indigenous hip-hop scenes in Nowra and the Torres Strait Islands (TSI). In these places, hip-hop becomes a form of glocalised creative expression, a means to personal development, and simultaneously a politicised, transnational and anti-colonial creative industry. Before we turn to these case studies, it is necessary to outline the wider discursive context within which Aboriginal hip-hop circulates—for as we shall see, this discursive context comes to be as important (if not more so) than physical remoteness or poverty in constraining and mediating successful growth of Indigenous hip-hop as vernacular creative industry.

Indigeneity and patron discourse

> There is no one kind of Aboriginal person or community. (Langton 1993, p. 12)

Within cultural policy and planning, Indigenous culture has often been assumed as static rather than as culture always in creation (Stavrias 2005; Morgan 2008). Dominant stereotypes—like those mobilised in tourism (Waitt 1999) and cultural industry promotion, especially visual art (Luckman *et al.* 2009)—have presumed Indigeneity as bounded by tradition, ethnicity and heritage. Rather than presume that 'Aboriginal' or 'Torres Strait Islander' is a racial, biological or natural concept, Marcia Langton argues that Indigeneity is a form of intersubjectivity, 'in that it [Indigeneity] is re-made over and over in a process of dialogue, imagination, representation and interpretation' (Langton 1993, pp. 33–4).

Biological constructs of Indigeneity have long been shrouded in prejudiced ideas regarding racial purity—related to skin colour—and 'blood' inheritance. Opposing racial concepts, the idea of Aboriginality as intersubjectivity aspires to dismantle discriminatory representations and empower Aboriginal and Torres Strait Islanders to negotiate and resist racial or biological perceptions. Media forms including TV, film, literature and music are important arenas for understanding how Aboriginality is intersubjectively constructed by both Aboriginal and non-Aboriginal people (Michaels 1986; Langton 1993; Dunbar-Hall & Gibson 2004).

While Indigeneity should be understood as intersubjectivity, it is also arbitrated by geography and industry (Gibson & Dunbar-Hall 2004). At one level (and particularly for Indigenous cultural producers), identity is connected to perceptions of place and space; cultural identifications are recognised as spatially dependent (Brubaker & Cooper 2000). Place is crucial; sites of creation and birth, ceremony, celebration, historical and spiritual, all contribute to the constructions of individual and group categories of identification. At another level, industries that sell Indigenous creativity frequently trade on constructions of Indigenous identity. This has profound implications for emerging Indigenous cultural industries, and their success (or otherwise) in markets beyond Indigenous audiences. An example of the way in which Indigenous culture and creativity have been framed within essentialist discourse is in Aboriginal art from the Northern Territory. Traditional dot or x-ray art painted by artists from remote communities is considered 'proper' Aboriginal art, while authentic Aboriginal music is said to contain elements of 'traditional' culture, such as the didjeridu, or at least blend contemporary styles with traditional sounds (Gibson 2006). Viewed as 'traditional', these activities are promoted to tourists and visitors to northern Australia, while more contemporary or avant-garde activities escape touristic and commercial representation and are less often considered 'authentically' Aboriginal.

Colonial assumptions of Aboriginal and Indigenous cultural identity are inhibited, within what Langton (1993, p. 27) describes as 'an ancient and universal feature of racism: the assumption of the undifferentiated other'. This requires Aboriginal and other minorities to perform or be creative within symbolic frames, installed and maintained by demands of Western audiences for 'authenticity' (Dunbar-Hall & Gibson 2004). Van Toorn (1990, pp. 102–3) calls these 'patron discourses', invoking sets of normative expectations within colonial society about how minority texts and artists should look, listen, speak and perform:

> To address an audience is to hold it in one's power; but it is also to place oneself in its power, to expose oneself to its judgements, its categories, the rules and customs that pertain in its culture ... The speech, writing and other cultural practices of minority groups are only liberated into the public domain to the extent that patron discourses succeed in trapping them in the categories which the dominant audience has available to contain them.

Against a backdrop of various attempts by Indigenous hip-hoppers in Nowra and the TSI to negotiate physical remoteness and socio-economic deprivation, we also explore below how patron discourses both provide and constrain performance opportunities.

Research contexts: Nowra and the Torres Strait Islands

Nowra, the main town in the Shoalhaven Local Government Area (LGA) on the south coast of NSW (see Figure 1), has a resident population of 30 000. During the summer holidays the town bulges as tourists pass through in accessing holiday spots in nearby Jervis Bay, Sussex Inlet and Ulladulla. Yet seasonal tourist traffic has brought Nowra limited economic or social benefit. Being inland, it has shared less in the tourism boom, instead constituting a regional service and retail centre along the major highway. It is a place characterised by high youth unemployment levels, out-migration and welfare dependence (ABS 2009a). Some 6 per cent of the population identify as Indigenous (ABS 2009a), a high proportion compared with national and State figures (2.1 per cent and 2.3 per cent, respectively), with 1 in 5 Indigenous people in Nowra aged 15–24 (ABS 2009a). In formal education, only 16 per cent of Indigenous youth complete a year 12 education in Nowra, compared to 31 per cent for non-Indigenous youth (contrasted to the national average of 42 per cent). Unemployment amongst the Indigenous population is 22 per cent, with a third of Indigenous youth aged 15–24 unemployed.

Another demographic feature of Nowra is the 'substantial net out-migration among those of school leaving age' (Shoalhaven City Council 2005). Between 2001

FIGURE 1. Location of Nowra, NSW and the Torres Strait Islands, QLD, Australia.
Source: Reproduced with permission from Geosciences Australia (2010).

and 2006 more than 10 per cent of Nowra's youth moved away, seeking employment and a better quality of life. Sydney, located 170 km to the north, is too far to commute to work or a social night out, so the best option often becomes relocation. Positioned at the margins of economic growth and social life, Nowra is a town facing complex problems, retaining a reputation for racial tensions, high rates of crime and violence, particularly in East Nowra, where around 20 per cent of the population is Indigenous (ABS 2009a; Shoalhaven City Council 2009). Stigmas are commonly attached to Indigenous youth in the town, depicted as delinquent, idle, and troublesome (cf. Dufty 2009). It is against this socio-economic and discursive background that we explore hip-hop in Nowra as vernacular Indigenous creativity.

The TSI, located off the coast of Far North Queensland, could perhaps be no more different to Nowra: home to 9000 residents scattered across 17 inhabited tropical islands. More than 80 per cent of the population in the TSI identify as Indigenous, related ethnically to Melanesia further to the north rather than to mainland Aboriginal nations. The dominant spoken languages are TSI Creole and other traditional Island dialects. Overall, the TSI comprise around 11 per cent of Australia's Indigenous population, but only 14 per cent of Islanders still live in the Islands (Trewin & Madden 2005). The physical distance of the TSI from the 'rest' of Australia cannot be easily overstated. Thursday Island (TI), the main population centre, is more than 3500 km from Melbourne and 2000 km from Brisbane. Where links exist with the mainland, they are commonly with Cairns (because of air transport) in Far North Queensland. Remoteness is a tangible part of everyday life in the TSI.

Although contrasting wildly in cultural and geographical terms, socio-economic trends in the TSI share some similarities with Nowra. The Indigenous population is four times more likely to be unemployed and more than twice as likely to be living in a low-income household compared to the non-Indigenous TSI population (ABS 2009b). Moreover, the retention rate in formal education for Indigenous youth in the Islands is half that of non-Indigenous youth; a similar ratio holds for non-school qualifications, such as a trade or diploma (Trewin & Madden 2005).

In these two socio-economically disadvantaged and geographically marginal settings flourishing Indigenous hip-hop scenes have emerged, overwhelmingly dominated by young people. Here, hip-hop is an example of creativity inspired by transnational cultural flows (and attendant linguistic and political features), but forming and operating within local spaces, geographically removed and socio-economically isolated from prosperous cities (places more predictably featuring in creative cities research). Creative city ideas developed elsewhere—of critical mass, cluster effects and creative milieu, considered key for urban creativity and innovation—are tested deeply in these case studies.

Methods and tools: researching Indigenous hip-hop

Our research approach sought depth of insights, but needed to remain flexible. The project initially focused on Nowra, concerned with 'hanging-out' and meeting young people involved in music. The local youth centre became a key research location, regularly utilised by youth participating in hip-hop. Hanging out at the centre helped build friendships and trust with a number of young hip-hoppers. After initial meetings, participant observation and a research diary were used to reflect on meetings, both *in situ* and out of context (Kearns 2005). These notes then formed the basis for subsequent semi-structured interviews. Six young rappers were interviewed, but our focus here is on the music produced by one hip-hop group who call themselves Yuin Soldiers, and in particular their three rappers Yung Nooky, Nat and Selway (see Plates 1 and 2). Combining interviews and participant observation with the group's music making provided a rich and extensive outline of processes for creating Yuin Soldiers' beats, lyrics and performances.

Next, the research drew on personal networks, expanding the focus to incorporate hip-hop from the TSI. Knowledge of the growing 'scene' in remote northern Australia allowed researchers to access a hip-hop crew called 'One Blood Hidden Image'. The group consists of six members from across the TSI. Our interviews and subsequent conversations were conducted with four members of the group; Maupower, Mondae, Cagney and Big Worm. We also listened closely to their songs and performances, which had been uploaded onto YouTube.

Methodologies varied for each study location. For Nowra, time could be spent moving through the youth centre, observing interactions and performances taking place within its spaces. A more ethnographic, in-depth analysis could therefore be undertaken, with data collection drawn out from extended visits with participants. For the TSI, methodologies needed to be more flexible. It was not possible to visit islands directly; instead, in-depth phone conversations, interviews and e-mailing with our TSI participants took priority.

Analysis of interviews, research notes and music then drew on an adapted form of narrative analysis. Narrative analysis is argued by Wiles *et al.* (2005) to be a sensitive way of writing fieldwork into geographic research 'because it focuses on how people talk about and evaluate places, experiences and situations, as well as what they say'—in our case what was said and rapped (cf. Skelton & Valentine 2005). One approach to analysis occurs where several narratives are used as 'case studies' to demonstrate different aspects of the same conceptual outcome (see Gorman-Murray 2006). In this case, differences are brought out across narratives to construct a range of emergent 'themes' reinforcing the same point, from different perspectives. Following this approach, analysis needed to be sensitive to individuals and their hip-hop stories, providing opportunity to acknowledge how each respondent built up their hip-hop experiences, ideas, networks and knowledge.

PLATE 1. Yung Nooky in the studio at Nowra youth centre. *Source*: Authors.

Plate 2. Yung Nooky and Selway performing at an Indigenous celebration event. *Source*: Authors.

Producing Indigenous hip-hop

Torres Strait: One Blood Hidden Image (OBHI)

Hip-hop music has become very popular among Indigenous youth in the TSI, attracted to its fast, funky beats, expression and accessibility. Big Worm, the newest OBHI member explained its popularity:

> We were doing it on the streets, around the Straits, fucking around, then Patrick came out with that single 'Home boys', that's when we all got like, yeah we can all do that too you know. That's when we started getting into it, we loved it, bro. Its poetry, like what you go through [in life], it's a good opportunity to use that in hip-hop. Now we be walking around town and stuff, and these younger fellas start coming up everywhere rapping, and we like, yeah you're good man, keep it up. (Big Worm)

OBHI have gained increasing exposure within the Indigenous hip-hop scene (see Plate 3). Comprising originally five main MCs—Patrick aka Maupower, Josh aka Cagney, Damien aka Mondae, Dayne aka Dayne-Jah, and Leroy aka Artu, the group have since incorporated several other members, including Troy aka Big Worm. OBHI work to create a distinctive hip-hop sound, mixing traditional Creole language with cultural stories and messages. The music has appealed to Indigenous elders across the TSI, who recognise its widespread appeal for youth. According to Maupower:

> It's Torres Strait Island hip-hop, an Indigenous hip-hop, we incorporate our language and culture into that style, that genre ... We get a great response from the elders cause that's a new genre for them. They're not used to hip-hop, and we show that we can incorporate our culture into hip-hop, and their like WOW, keep it up. (Maupower)

To create and produce their own unique beats, sounds and rhymes, OBHI have become skilled at using computer music programs such as Fruity Loops, Acid and Reason that give aspiring hip-hoppers in remote locations the ability to sample and

PLATE 3. OBHI's *Shut the Gate* album (2008), sold through the group's website. *Source*: www.maupower.com.au

fuse together sounds to compose original beats, without relying on city recording studios:

> Well I started using Fruity Loops, now we have upgraded. I'm using Adobe Edition, a bit of Reason, ACID. I use Reason for the samples but I still record [raps] in Adobe. (Maupower)

While all performers in the group created their own beats using computer programs, Mondae was the 'lead beat master':

> I listen to the music, focus on the beats. It comes natural, some days I can do 3 beats. I use the laptop and Fruity Loops, go through every instrument [in the program], mix in different instruments, change up

149

the pace, go through, clean it up, til you've got something ya like. (Mondae)

Making beats was a technical skill, requiring practice and refinement. Computer programs were cheaper than DJ equipment and could be self-learnt, allowing participants to sample instrumental segments from other genres. OBHI liked to 'cut' in reggae sounds, snare drums, with deep bass guitars, manoeuvring the pace of beats, slowing down or speeding up, depending on message and the type of song. After the beat and rap were brought together the group recorded their music using Maupower's homemade studio. Emerging technologies of hip-hop production created a more accessible, do-it-yourself (DIY) musical form. The order for creating a new hip-hop song was variable for OBHI, dependent on time, the availability of members and motivation:

> You know like sometimes, we have a beat and we write the lyrics, sometimes we have the lyrics and mix and compose a beat to it, depends how we feel at the time. Sometimes we just get into the studio and do it all up there at the same time, on the spot. (Cagney)

OBHI uploaded music onto video networking sites like MySpace and YouTube, where two tracks, 'Coolies' and '4 the Balaz' have received 20 000 'hits', a significant number for an underground, unsigned group. The band also sells and promotes their music (four albums) online via their website, along with clothing and other merchandise (see www.maupower.com.au). Their sound is driven and produced by modern technologies and techniques, mostly circulated electronically. Group members are often stopped in the street and praised for their music, while Maupower was nominated for a 2009 Deadly Award (the national Indigenous music awards). Through beat mixing, rhyming, performance, dancing, and computer skills, hip-hop is a means to be creative.

Nowra: Yuin Soldiers

On most afternoons the Nowra youth centre's music rooms are occupied by groups of young hip-hoppers. Emerging here is a group called Yuin Soldiers, who have a shifting line-up including Corey aka Yung Nooky, Nat and Selway, Nooky's cousin. Recognising a growing interest by young people from the area in hip-hop, the youth centre has built two music rooms used by budding musicians for learning, practice, and performing. The soundproof room allows hip-hoppers to mix beats and rhymes, record their tracks and place them onto CD or Mp3 player. The services at the youth centre are crucial for grassroots music making in Nowra, providing the only free space in town to use dee-jaying turntables, mixing, editing and recording equipment.

The production of new songs for Yuin Soldiers, like for OBHI, was reliant on computer programs and technologies, creating sounds and beats for the rappers. For Nooky, creating 'cool' beats was a skill that required practice and perseverance. He credited Selway for helping with beat creation:

> Sometimes I don't pick it up, and I get frustrated at it. I'm getting better, my cousin [Selway] helps the most. He can just rip em out. Too easy. And they sound so good. It's a skill that I'm learning. (Nooky)

Nooky outlined how he composed songs:

> ... We put the beat on there first [demonstrates on the computer screen], then we rap to the beat. You do your back-up vocals and you compress it, bounce it down and it's ready for CD. Sometimes I can do it in 40 minutes, but then sometimes it can take a few hours or days to do a song. It depends. I'm always writing. When I'm at home I write, and when I'm at school ... (Nooky)

To assist Yuin Soldiers' hip-hop, more established Indigenous artists provided encouragement, informal schooling and even direct help with composing sounds and beats. Older performers like Wire MC, Brotha Black and Street Warriors were seen as 'inspiration' for Yuin Soldiers and their music making:

> There's Wire MC, I think he's related somewhere down the line ... he's really good, Brotha Black and Street Warriors, and all the Indigenous rappers they inspire me. (Nooky)

Wire MC and Indigenous performer Choo Choo (CuzCo) had previously collaborated with Yuin Soldiers. For Nooky, his older cousin Selway, a skilled hip-hopper from a group called East Coast Productions (ECP), was another prominent figure assisting in his musical production:

> I usually get my beats off my cousin [Selway] because mine still aren't that good yet [laughs], he gives me a lot of beats and we sit at home and sometimes I think of stuff and start writing or something happens and I'll just start writing about it. If I write something than I'll just ask my cousin for a beat. (Nooky)

OBHI, with most members in their 20s, had been practising and refining hip-hop over several years. Yuin Soldiers—Nooky, Selway and Nat—aged in their late teens, were younger and less experienced performers. They spoke of building up skills for music making. In the same way as OBHI were promoting hip-hop in the TSI to younger, budding hip-hoppers, 'showing them the way' (Big Worm), Yuin Soldiers had drawn on the expertise and experience of more established acts like Wire MC and Choo Choo, to advance their creative skills.

Managing and navigating remoteness through music

Torres Strait Islands: OBHI

Each member of OBHI was born on different parts of the TSI. This has geographic significance for the group, for, as Maupower explained, OBHI invented their name through their geography:

> We were singin like nobody, we didn't have a name; just called the rap group. The original five members were sitting around and said we want to come up with a name which could be a concept that represented us as a group. We tried to find one, because each member represents one particular region. The Torres Strait is subdivided into 5 different regions; we have the inner islands, the near western, the central, the top western and the eastern islands, and each member came from that blood line, those regions and we were all related. So we had a blood line connection,

and so we said we are all one blood, with no particular image, so we all had different forms, and that's how we evolved to One Blood Hidden Image. (Maupower)

Rather than isolation and remoteness from large cities being paralysing for creativity, OBHI overcame distance through hip-hop. Combining music with new technologies, like YouTube, the TSI are positioned in their hip-hop as a musical hub. Indeed, the group have uncovered opportunities to travel and experience the rest of Australia, performing their music, in Brisbane, Sydney, Newcastle, Melbourne, Adelaide and Auckland. Their songs project messages about brotherhood and maintaining a positive outlook:

> Blood is what you make it, how else can I say it?
> Who has got your back when the end of the day hits?
> Different day same shit? NO, I wouldn't change it,
> [chorus] This is something for my Balaz [brothers],
> Through the struggles that be holding you down,
> Keep your head up, you gotta keep your head up.

> (OBHI- '4 the Balaz' (brothers): see http://www.youtube.com/watch?v=UhantuJZpBY)

Cagney and Maupower explained how isolation and remoteness were managed in the TSI. As the established hip-hoppers in the region, the group became involved in schooling younger budding rappers, especially in their own hip-hop production, giving the 'young fellas' something positive to do:

> Well making it up here is easy, I show most of them [young fellas] how to build a home studio and record it, we all use the same program, so we'll all interact together. (Maupower)

> There's a lot of mob now starting to come out. There was a lot of shame. Shame was big up here, and so for them young kids when we up on stage, we say come down here, watch us, we notice how all of a sudden they have courage to get up themselves. We do workshops too; do up a beat and each person has a line by line. Everybody got their story and in this way [hip-hop] even the smallest voice can be heard you know. (Cagney)

Overcoming the 'shame' factor, performing and expressing themselves in front of audiences was an issue facing many young people in the TSI. Those who had taken the 'jump' forward to performing their music had gained important benefits, according to Big Worm:

> When they have a go at the workshop, rap to the beat, the young fellas go yeah, this is cool. We get them to write their raps down, then they can record and play them back. It gives them a buzz bro eh? Like you see it on their faces; Fuck, we can do it. (Big Worm)

Creating hip-hop was accessible for most young fans in the TSI because of cheap technology and available mentoring, giving them a 'positive thing to do' (Cagney). Maupower took an active role in schooling younger enthusiasts, showing youth how to set up their own recording areas in the home or garage. The local library and TAFE also provided places to practise music. While remote and marginal, the TSI has a growing music scene with creativity funnelled into the production of beats and rhymes.

Nowra: Yuin Soldiers

While not as geographically remote as the TSI, Nowra is socio-economically and in a certain way also geographically marginal. This marginality is openly confronted by Yung Nooky and Nat, through their raps and narratives, performing both individually and as Yuin Soldiers with Selway. When asked about the origins of their group's name, Nooky explained that:

> Yuin is our people, like where we come from, and soldiers, they keep fighting and never give up, so that's where the name Yuin soldiers came from. (Nooky)

Their music confronts prejudiced experience, with Nowra seen as a place perpetuating racialised ideas of Aboriginality. One of Yuin Soldiers' songs 'Subliminal twist' raps about the marginality experienced by Indigenous youth in Nowra:

> Blackfella on the hunt,
> Sick of being called a little black cunt,
> While I'm walking I'm thinking,
> Is this the price of education?
> Heartache, racism and discrimination?
> I'm sick of teachers saying these kids ain't black,
> Just because were not as dark as them Williams' girls ...
> [Chorus]
> You had your chance and you couldn't make me quit,
> 3 months from now, I'll be done with this shit,
> So until then I'm a stay strong and continue to spit,
> These lyrics with a, with a subliminal twist, subliminal twist.
>   (Yuin Soldiers)

Nowra was consistently identified as a place which embodied intolerance. Yuin Soldiers' music was a powerful way of breaking down and confronting these issues. Also, in parallel to their sense of discouragement, were feelings of attachment to Nowra and the south coast:

> Well Nowra is where we live and grew up so its home, that's a strong feeling, like this is your place. But it's also a place that gives you the shits. You've got to get out of Nowra for a while; it can get you down, but go away, then come back and keep goin. It is a beautiful place and that, but it can be a pretty racist place you know? (Nooky)

Nowra is home, yet consciously is also a place to escape. Living in Nowra is seen as a struggle or fight for Indigenous youth, metaphorically drawn out in a Yuin Soldiers rap, where Nooky makes comparison with American boxer Ruben 'Hurricane' Carter:

> South Coast Hurricane ... you can call me Ruben Carter,
> Instead of a right hook, it's the rapper Yung Nook ...

The first round's already won,
2541 ask around I'm the man in that town,
I'm goin big with my South coast sound.
[Chorus- Nooky and Nat]
Ah welcome you all to the South coast flow,
On the map we're big, that's how we roll, yeah got the endless rhymes,
yeah for the endless crimes.

In Nowra's Aboriginal hip-hop, local experience is integrated with a politicised transnational black culture. Music making can be drawn from daily experiences within marginal places, providing creative stimulus for rhymes and raps. Hip-hop allows negotiation through confrontation, using a 'loud, cool and stylish' (Nat) form of music.

Performing hip-hop

Torres Strait Islands: OBHI

While practising in the TSI was considered 'easy', gaining access to performance spaces outside the islands was more difficult and attributed to the region's 'remoteness' (Mondae). However, from analysis of interview transcripts and research diary notes, marginality emerged as not the only barrier to performance as OBHI sought widespread recognition. Gigs have been dominantly bounded within Indigenous ceremonies, events or festivals: National Sorry Day, Reconciliation events and NAIDOC celebrations. These performances provided rare travel experiences to locations across mainland Australia; hence they were recounted very positively.

Yet Maupower, Mondae, Big Worm and Cagney said in interviews how remoteness in the TSI left musically talented youth frustrated, unable to access opportunities to play their music to larger audiences on the mainland, or make any significant income from their work. OBHI themselves rarely received invitations to play at music festivals or gigs outside of symbolic Indigenous events:

> Well we started out here performing in NAIDOC, cultural weeks up here, cultural festivals up the Torres straits, and we started moving down to like the styling up festivals [Indigenous hip-hop event] in Brisbane, Survival day in Adelaide, and Sydney as well, but we perform mostly cultural events. (Maupower)

To improve musical skills, participants spoke of 'moving to the mainland' (Big Worm) for education and opportunity. On the mainland Maupower and Mondae had learnt more 'formal' music skills—recording, computer programming and professional networking—called upon for accessing performance opportunities. Access thus required participants to move outside the Islands. Big Worm told of needing to 'get away' from the TSI, recently deciding to move to Brisbane for greater opportunities:

> When I was living there, it was no opportunity; I want to do more music. I've got to know a few of the artists around here, and you say can I come on your track, so you try and get an opportunity like that but it's hard. We did play Styling Up in Brisbane, in Sydney for an anti-government

thing ... We try to get gigs paid for and accommodation paid for, but some people just don't pay that money, so we all have to chip in money to do it. You do it for the passion. (Big Worm)

Playing gigs to non-Indigenous audiences was rare; receiving payment for their shows or funding for recording or workshops was rarer still. Members had strong aspirations to forge professional music careers, moving away from the TSI, at various times, to pursue those goals. However, their ambitions of becoming professional hip-hoppers were yet to materialise, in part due to patron discourses restricting avenues for paid performance.

Nowra: Yuin Soldiers

Yuin Soldiers also commented on the limited support for musical performance in Nowra. Indigenous celebrations such as NAIDOC events supplied the majority of their hip-hop performance opportunities. In addition to Yuin Soldiers' hip-hop, Nooky and Selway are involved in traditional Aboriginal dancing. Most performances outside of Nowra privileged their traditional Aboriginal dancing over hip-hop. It was rare for any of the young rappers to play a hip-hop gig at school for a band or music day, or at non-Indigenous music festivals. Music celebrated at these events was likely to be other genres, like punk, rock music, or so-called traditional Aboriginal performances, like didjeridu playing or dancing:

Mostly, like the NAIDOC week people ringing me up, and like [pause] yeah there's places I have performed at, like here [Nowra youth centre], and I'm performing here next Friday, and up at the showground in the middle of town but you mostly get booked out for Indigenous type events. (Nooky)

On occasions when Yuin Soldiers' performed outside Indigenous events, participation was often confined within touristic representations of Aboriginality, confined to traditional dancing. Nooky and his cousin Selway had performed traditional Aboriginal dances for World Youth Day in Sydney, in front of a large audience, before being invited to perform a hip-hop set, which fused traditional dancing elements with their beats and rhymes. This was a highlight, lamented as a rare opportunity to showcase their contemporary rhyming and performance skills. The lack of hip-hop opportunities contrasted with the praise from audience members and other more established musicians, at their gigs. A local Aboriginal community event, called 'A new beginning', celebrated during reconciliation week, was a performance that received special acclaim from an unlikely fan:

Yeah like at that Bundanoon one [gig], this old, this lady came up and said 'I don't like rap but I love what you did'. And this fella from Wollongong Mr. McFlawless, yeah he got in contact with us and said it was good, and [I] went up [to his house] and did a song with him. (Nooky)

Despite performing a modern, creative and expressive Aboriginality ('our culture, it's like, we sing and dance and hip-hop fits in good'—Nat), there was a scarcity of encouragement from the wider non-Indigenous community. With ambitions to become a professional performer, Nooky, like Maupower and Big Worm, has relocated from his local town, enrolling in a formal performing arts school on the

NSW central coast, hoping his relocation will help access career opportunities in performance.

Conclusions

Aboriginal hip-hop is 'creative' because it is concerned with being artistic, resourceful and innovative. Hip-hop is a glocal subcultural performance (Mitchell 2003), and an oppositional musical form (Iveson 1997). Importantly, creativity is central to its practice. Music making involves acts of craftmanship, manoeuvring, recording and performing.

Hip-hop is also a fusion between the traditional (language, cultural stories, histories and dance) and contemporary (equipment, software and technologies), music richly performed by Indigenous participants, appropriated through transnational black networks, across diverse locations. Despite isolation from centres with large populations, creative buzz and critical mass—the factors seen as crucial in much creative industries literature for fostering and promoting creative talent— Nowra and the TSI emerge as hubs for the creative production of this contemporary music form. Their local hip-hop scenes highlight the possibility for creativity, richly present outside creative cities.

Certainly, new media and communication technologies have increased the accessibility of contemporary music forms, shifting exchanges of information and symbolic goods (Burgess 2006; Kruse 2009). Production and consumption of music is increasingly reliant on these emerging technologies. Individual songs, albums and entire discographies are downloaded from Internet sources such as iTunes or Limewire in minutes, transferring music to CD and Mp3 players. Computer programs help develop unique sounds and beats, often replacing the need to learn technical instruments. Music can be made at home or at the local youth centre, recorded and uploaded to YouTube or Facebook sites. In turn these networks spread music to larger audiences, promoting skills and ambitions of grassroots, underground, as well as signed professional performers. Key to it all is a sense of solidarity among hip-hoppers (both Indigenous and non-Indigenous) and a desire to expand the creative community through sharing, learning and helping others.

When talented, aspirational young rappers demonstrate creative skill and ability, geography has not been the only barrier to success or opportunity. Patron discourses have restricted performance opportunities. Both OBHI and Yuin Soldiers have to navigate wider expectations. These expectations have framed possibilities, at times creating but also constraining opportunities. Like many other 'hidden' hip-hoppers across Australia, OBHI and Yuin Soldiers articulate an ambition to integrate creativity into professional, paid or career work. They have identified their talents and skills and have been applauded for them, but face locational and discursive barriers for implementing a pathway into professional creative work.

Acknowledgements

This research was funded through the Australian Research Council's Linkage Project scheme (LP0882238).

NOTE

[1] Rob Evitt is Indigenous and belongs to Yirandali Aboriginal nation, in the Hughenden area of north-west Queensland. Andrew Warren is non-Indigenous, completing his PhD in Human Geography at UOW.

REFERENCES

AUSTRALIAN BUREAU OF STATISTICS (ABS) (2009a) 'Community profiles, Nowra', available from: www.abs.gov.au (accessed 16 August 2009).

AUSTRALIAN BUREAU OF STATISTICS (ABS) (2009b) 'Community profiles, Torres Strait Islands', available from: www.abs.gov.au (accessed 12 September 2009).

BRADLEY, L. (2000) *Bass culture: when Reggae was king*, Viking, London.

BRUBAKER, R. & COOPER, F. (2000) 'Beyond identity', *Theory and Society* 29(1), pp. 1–47.

BURARRWANGA, L., MAYMURU, D., GANAMBARR, R., GANAMBARR, B., WRIGHT, S., SUCHET-PEARSON, S. & LLOYD, K. (2008) *Weaving lives together at Bawaka, North East Arnhem Land*, CURS, University of Newcastle, Newcastle.

BURGESS, J. (2006) 'Hearing ordinary voices: cultural studies, vernacular creativity and digital storytelling', *Continuum: Journal of Media & Cultural Studies* 20(2), pp. 201–14.

DUFTY, R. (2009) '"At least I don't live in Vegemite Valley": racism and rural public housing spaces', *Australian Geographer* 40(4), pp. 429–49.

DUNBAR-HALL, P. & GIBSON, C. (2004) *Deadly sounds, deadly places: contemporary Aboriginal music in Australia*, UNSW Press, Sydney.

FARLEY, C.J., AUGUST, M., BRICE, L.E., HARRISON, L., MURPHY, T. & THIGPEN, D.E. (1999) 'Music: hip-hop nation', *TIME magazine*, available from: http://www.time.com/time/magazine/article/0,9171,990164,00.html (accessed 12 October 2009).

GEOSCIENCES AUSTRALIA (2010) 'Cartography: map of Australia with state capitals', Commonwealth of Australia, Canberra, available from: http://www.ga.gov.au/image_cache/GA5566.pdf (accessed 5 February 2010).

GIBSON, C. (2006) 'Decolonising the production of geographical knowledges? Reflections on research with Indigenous musicians', *Geografiska Annaler B: Human Geography* 88, pp. 277–84.

GIBSON, C. & CONNELL, J. (2004) 'Cultural industry production in remote places: Indigenous popular music in Australia', in Power, D. & Scott, A. (eds) *The cultural industries and the production of culture*, Routledge, London and New York, pp. 243–58.

GIBSON, C. & DUNBAR-HALL, P. (2004) 'Mediating Aboriginal music: discussions of the music industry in Australia', *Perfect Beat* 7(1), pp. 17–41.

GORMAN-MURRAY, A. (2006) 'Homeboys: uses of home by gay Australian men', *Social and Cultural Geography* 7(1), pp. 53–69.

IVESON, K. (1997) 'Partying, politics and getting paid: hip-hop and national identity in Australia', *Overland* 147, pp. 39–47.

KEARNS, R. (2005) 'Participant observation', in Hay, I. (ed.) *Qualitative methods in human geography*, Oxford University Press, Melbourne.

KITWANA, B. (2003) *The hip hop generation: young blacks and the crisis in African American culture*, Basic Civitas Books, New York.

KRUSE, H.C. (2009) 'Local independent music scenes and the implications of the internet', in Johansson, O. & Bell, T.L. (eds) *Sound, society and the geography of popular music*, Ashgate, Aldershot, pp. 205–17.

LANGTON, M. (1993) *Well I heard it on the radio and I saw it on the television: an essay for the Australian Film Commission on the politics and aesthetics of filmmaking by and about Aboriginal people and things*, Australian Film Commission, Sydney.

LUCKMAN, S., GIBSON, C. & LEA, T. (2009) 'Mosquitoes in the mix: how transferable is creative city thinking?', *Singapore Journal of Tropical Geography* 30, pp. 70–85.

MICHAELS, E. (1986) *The Aboriginal invention of television in central Australia*, Australian Institute of Aboriginal Studies, Canberra.

MITCHELL, T. (2003) 'Australian hip-hop as a subculture', *Youth Studies Australia* 22(2), pp. 1–13.

MITCHELL, T. (2006) 'Blackfellas rapping, breaking and writing: a short history of Aboriginal hip-hop', *Aboriginal History* 30(1), pp. 124–37.

MORGAN, G. (2008) 'Global subcultures and creative cities discourse', RC 21 conference, Tokyo.

PENNYCOOK, A. (2007) 'Language, localization and the real: hip-hop and the global spread of authenticity', *Journal of Language, Identity and Education* 6(2), pp. 101–15.

SAMY ALIM, H. (2007) 'Critical hip-hop language pedagogies: combat, consciousness, and the cultural politics of communication', *Journal of Language, Identity and Education* 6(2), pp. 161–76.

SHOALHAVEN CITY COUNCIL (2005) 'Economic development strategy', available from: www.shoalhaven.nsw.gov.au (accessed 16 August 2009).

SHOALHAVEN CITY COUNCIL (2009) 'Community crime prevention plan', available from: www.shoalhaven.nsw.gov.au (accessed 10 October 2009).

SKELTON, T. & VALENTINE, G. (2005) 'Exploring notions of masculinity and fatherhood: when gay sons come out to heterosexual fathers', in Van Hoven, B. & Horschelmann, K. (eds) *Spaces of masculinities*, Routledge, London, pp. 207–21.

STAVRIAS, G. (2005) 'Droppin' conscious beats and flows: Aboriginal hip-hop and youth identity', *Australian Aboriginal Studies* 2(1), pp. 44–54.

TOOP, D. (1984) *The rap attack*, Pluto Press, London.

TREWIN, D. & MADDEN, R. (2005) 'The health and welfare of Australia's Aboriginal and Torres Strait Islanders', Australian Bureau of Statistics, available from: http://www.aihw.gov.au/publications/ihw/hwaatsip05/hwaatsip05.pdf (accessed 16 August 2009).

VAN TOORN, P. (1990) 'Discourse/patron discourse: how minority texts command the attention of majority audiences', *SPAN* 30(1), pp. 102–15.

WAITT, G. (1999) 'Naturalizing the "primitive": a critique of marketing Australia's Indigenous people as "hunter–gatherers"', *Tourism Geographies* 1(2), pp. 142–63.

WHITE, C. (2009) 'Rapper on a rampage: theorising the political significance of Aboriginal Australian hip-hop and reggae', *Transforming Cultures e-Journal* 4(1), available from: http://epress.lib.uts.edu.au/journals/TfC (accessed 14 September 2009).

WILES, J., ROSENBERG, M. & KEARNS, R. (2005) 'Narrative analysis as a strategy for understanding interview talk in geographic research', *Area* 37(1), pp. 89–99.

Index

Page numbers in *Italics* represent tables.
Page numbers in **Bold** represent figures.